Village Voices, Forest Choices

Joint Forest Management in India

*To Toby, Nathan, and
all the world's forest children*

Village Voices, Forest Choices

Joint Forest Management in India

edited by

MARK POFFENBERGER

BETSY McGEAN

DELHI

OXFORD UNIVERSITY PRESS

BOMBAY   CALCUTTA   MADRAS

1996

Oxford University Press, Walton Street, Oxford OX2 6DP

Oxford New York
Athens Auckland Bangkok Bombay
Calcutta Cape Town Dar es Salaam Delhi
Florence Hong Kong Istanbul Karachi
Kuala Lumpur Madras Madrid Melbourne
Mexico City Nairobi Paris Singapore
Taipei Tokyo Toronto
and associates in
Berlin Ibadan

ISBN 0 19 563683 X

A collaborative effort of the
Asia Sustainable Forest Management Network
and the
Society for the Promotion of Wastelands Development
with support from
East West Center ■ Ford Foundation ■ University of California, Berkeley

Typeset by Rastrixi, New Delhi 110070
Printed in India at Pauls Press, New Delhi 110020
and published by Neil O'Brien, Oxford University Press
YMCA Library Building, Jai Singh Road, New Delhi 110001

Foreword

This is a book of contemporary concern narrating the emergence of grassroots forest movements in India. The book uses a historical perspective to assess the implications of shifting forest management policies and practices. In the past, people knew about their symbiotic relationship with forests and there was greater harmony between man and nature. More recently, commercial and population pressures have used the forest beyond its carrying capacity resulting in degradation. The book demonstrates that rural people are concerned over the disastrous consequences of overexploitation, including extensive soil erosion, falling ground water levels, and increases in arid micro-climates adversely affecting agricultural production and aggravating poverty. Driven by necessity, these villagers have spontaneously demonstrated their ability to protect and to restore badly degraded waste lands to productive healthy natural forests. In the past decade over 10,000 villages in the states of Orissa, West Bengal and Bihar have begun protecting and regenerating their natural *sal* forests, mostly unassisted by World Bank loans, bilateral grants or government projects. These predominantly tribal communities protect adjoining forests through volunteer patrols, armed only with staffs, and bows and arrows. Village men and women prevent tree felling, grazing, and fires, and in return sustainably harvest fuel, fodder, and foods. This is based on an age old tested philosophy of caring and sharing.

The book relates recent experiences with community forestry in India, setting them in an historical context, while suggesting their importance for the future. The importance of lessons and experience conveyed in this timely volume goes beyond India. The path our nation is following in empowering communities as caretakers of public lands will likely be one that is also taken by other developing nations in coming decades, and to some extent the industrialized world.

The book is targeted at a broad audience, written in a general, non-academic style, it speaks to students and professionals interested in both the social and ecological dynamics and challenges facing natural forest management today and in the future. Many experienced foresters, scientists and non-governmental organization (NGO) leaders contributed to the ideas in the book. It will be important for the tens of thousands of officers within the IFS and state cadres to read it and gain a broader understanding of the concepts, and strategies which Joint Forest Management entails.

I think one of the most fascinating aspects of the Indian experience with community forestry, is that the stabilization of natural forests is being achieved by some of the nation's poorest communities, in severely biologically disturbed forest areas. After several decades of mixed results from well-financed, technically sophisticated donor and government driven development programmes, the book suggests that there are indigenous, need-based, people-driven, low-cost ways to make some of India's forests healthy and productive again. Emerging local initiatives have the potential to be both socially just and ecologically sound. To succeed, we will need to listen to the *village voices* and to empower communities in *forest choices*.

Inspector General of Forests   M.F. AHMED
Government of India

Preface

The debate over the roles that communities and the state should play in managing forest resources is an ancient one. For more than 200 years, since the inception of British colonial power in India, politicians, business people, government foresters, and villagers have sparred over the issue. The dispute defies any absolute resolution. The desired path is to achieve a balance of tenurial rights and responsibility. Governments are powerful and enjoy the political, legislative, judicial, and financial strength to exert their will. Rural communities have numeric superiority, a strong, vested dependency, and a strategic geographical position in relation to the resources. Ultimately, governments rise and plummet, but the forest people, with their intuitive connections to the land, remain. Amid India's vast populace, the urgent livelihood needs of such biomass-based rural communities have motivated them to take action and assert greater control over their forests.

In the late 1980s, a group of twenty foresters, social scientists, ecologists, planners, and development specialists from South East Asia and the West wrote a book entitled *Keepers of the Forest*. The authors drew upon the historical record, as well as a wide range of community forestry field experiences from Indonesia, Thailand, and the Philippines. The objectives were to better understand the larger socio-political forces driving deforestation and to identify ways in which socially and ecologically sound systems of forest management could be established. Representing some of the most enlightened practitioners in the field, the group concluded by unequivocally advocating the empowerment of communities to serve as 'keepers' and primary beneficiaries of these important forest resources.

Five years later, concerned individuals of a diverse mix have combined their knowledge to examine the condition of India's forests and the prospects for improved productivity and management. The findings corroborate and build upon insights generated by South East Asian colleagues, and offer new perspectives and strategies towards answering the questions of both ecological and organizational sustainability in managing natural forests.

Acknowledgement needs to be given to the hundreds of Indian and British social reformers who struggled before and after Independence to recognize the importance of tribal peoples and forest communities as resource managers. The historical record of forest rights alienation has been extremely well documented by Indian scholars over the past 50 years. Their careful research facilitates our understanding of the roots of the forest management crisis and offers insights into its resolution. While it is impossible to express our gratitude to all those individuals who have contributed to the ideas gathered in this volume we would like to mention a few whose thoughtfulness and commitment over the years has been particularly influential. They include Kamla Chowdhry, Samar Singh, N.S. Jodha, K.C. Malhotra, Anil Agarwal, Sunita Narayan, Anil Shah, Robert Chambers, Ajay Mehta, Kartikeya Sarabhai, R.K. Patchauri, Shekhar Singh, Prabir Guhatakurta, Kundan and Neera Singh, Sashi Kant, Deep Pandey, Ram Sharma, J.R. Gupta, Syed Rizvi, N.C. Saxena, Tirath Gupta, S.K. Dhar, M.S. Swaminathan, B.M.S. Rathore, Raju and Manju, G.S. Mandal, P.K. Roy Choudhury, P.K. Das, I.Z. Khan, Narayan Banerjee, R.S. Pathan, and F.M. Ahmed.

It is difficult to pen acknowledgements adequately in a book to which so many have contributed. Village leaders, courageous men, women, and youth who have advocated, organized, and nurtured emerging community forest protection and management groups may be personally unknown to the authors, but their good work is apparent in many thousands of villages across India. The tribal communities of India — the Bhils and Vasava of south-eastern Gujarat, the Santhals and Lodhas of south-west Bengal, the Mundas and Gonds of Orissa — are only a few of the hundreds of tribals throughout India who intimately know and cherish the forest. Against its stark disappearance, these people are acting to save the forest for their survival and posterity. We admire and thank them for their heroic efforts, as we believe that they are the ultimate managers of India's forests. We are also grateful to the many Indian Forest Service officers who acknowledge the need for change in forest management policy and practice. Increasing numbers of foresters are showing their commitment to the fundamental problems underpinning mismanagement by experimenting with new solutions through a wholly new paradigm of partnership. They, too, must lead the way with resolve into the twenty-first century.

Scores of creative individuals from India's diverse array of non-governmental organizations (NGOs), large and small, have lobbied tirelessly for policy reforms and for the improved welfare of poor forest communities. They have developed new ways to facilitate communications between foresters and villagers, and have pioneered strategies to give rural women more authority in decision-making processes over natural resources affecting their lives. They have been the advocates for the disempowered when no other voices spoke for them. Each of these groups has contributed to a remarkable phenomenon taking place in India today — a reversal of the alienation of forest people's rights, of institutional conflict, and of ecological patterns of forest degradation. This transformation appears part of an important historical process. While it is still too early to predict whether this movement will reshape the complexion of India's forestry, it is significant that parallel patterns of community empowerment over local resources are emerging across a much wider latitude in developing and industrialized nations alike. Against the formidable odds of intense historical conflicts and severe current pressures on India's forest resource base, the fact that groups of NGOs, government agencies, and forest communities have increasingly begun to work together, jointly confronting management problems and devising practical, highly localized solutions, is most encouraging. The synergism of the collaborative action may well solidify and accelerate this process of change. Meanwhile, the resilience witnessed in nature's response is promising. Increasing field evidence in a range of India's forest ecotypes indicates that natural forests under effective community protection can regenerate rapidly into multi-tiered, biologically diverse secondary forests. For all of India's forest people, and the ecosystems that support them, we sincerely hope they will succeed in their co-operative venture.

Preparation of this anthology was generously supported by the Ford Foundation, the Wallace Genetics Foundation, the Asia Bureau of the US Agency for International Development, the USDA Forest Service International Forestry Program, and the USAID-funded Biodiversity Support Project of the World Wildlife Fund, the Nature Conservancy, and the World Resources

Institute. We are particularly grateful to Gordon Conway, Peter Geithner, Jeffrey Campbell, Molly Kux, George Taylor, Stephan Kelleher, Linda Lind, Alex Moad, A. Terry Rambo, and Jeff Fox for their encouragement. We are also indebted to the East-West Center Program on Environment for providing a writing workshop and crucial logistical support to the contributors of this volume. Thanks are also due to Helen Takeuchi for her thoughtful and fine-tuned editing of the chapters, and to Mary Hayano for her dedicated hours in front of the computer screen. Finally, we wish to thank the Delhi staff of Oxford University Press for bringing this book to our readers.

Contributors

AJIT BANERJEE, Consultant, The World Bank, Washington, D.C.

JEFF CAMPBELL, Program Officer, Ford Foundation, New Delhi

MITALI CHATTERJEE, Senior Associate, Indian Institute for Biosocial Research and Development

MADHAV GADGIL, Professor of Ecology, Centre for Ecological Studies, Indian Institute of Science, Bangalore

RAMACHANDRA GUHA, Visiting Fellow, Wissenschaftskolleg zu Berlin

DAVID HARDIMAN, Lecturer, School of Oriental and African Studies, University of London

ARVIND KHARE, Executive Director, Society for the Promotion of Wastelands Development

BETSY McGEAN, Social Ecologist, The World Bank, Washington, D.C.

SUBHABRATA PALIT, Chief Conservator for Social Forestry, West Bengal Forest Department

MARK POFFENBERGER, Director, Asian Sustainable Forest Management Network

N.H. RAVINDRANATH, Professor of Economics, Centre for Ecological Studies, Indian Institute of Science, Bangalore

S.B. ROY, Executive Director, Indian Institute for Biosocial Research and Development

MADHU SARIN, Consultant to the National Support Group on Joint Forest Management, Society for the Promotion of Wastelands Development

CHHATRAPATI SINGH, Director, Center for Environmental Law, World Wide Fund for Nature — India

Contents

Figures and Tables

Figures

Tables

Plates

Colour
Between pages 236 and 237

1. Men in uniforms with a paramilitary orientation continue to create social distance between foresters and villagers (M. Poffenberger).

2. In Bhainsara hamlet (*phalan*), 20 kilometres north of Udaipur, in Rajasthan, the Bhil hamlet chief says 'we have protected all the forest land on the far side of the ridge for ten years and all the trees and grasses are growing well. Two neighbouring villages are ready to join us. Sickle fees are charged for each family that cuts grass in the forest and we use the money to hire watchers from our village who guard the forest from outsiders who may try to fell our trees' (M. Poffenberger).

3. Using sketch maps Sikh, Jat and Banjara villagers work with J.R. Gupta of the Haryana Forest Department to clarify management responsibilities in Shivalik hill forests using sketch maps (M. Poffenberger).

4. A Vasava tribal woman in southern Gujarat shows foresters and village leaders, on a ground sketch map made of local materials, where she travels in the forest to collect mahua, gum, fodder and fuelwood (M. Poffenberger).

5. This map was commissioned by a forest protection committee in West Bengal. The villagers hired the local tax collector to draw the map to better clarify the boundaries of their protected forest (M. Poffenberger).

6. The removal of root stock is a final blow in the impoverishment of the forest ecosystem. Without these roots, topsoil erosion will accelerate and rapid regeneration possibilities are lost (M. Poffenberger).

7. Bhil woman from Bhainsara break for lunch after planting trees on degraded forest lands. Tens of millions of Indian women are primary forest users. New means must be found to bring them into management decision making (M. Poffenberger).

8. In arid Western Rajasthan, the indigenous Bishnoi people have survived for centuries, tapping an extensive knowledge of the desert and scrub forest environment. Their continued presence in the desert is ensured only by their sustainable management strategies (M. Poffenberger).

Abbreviations

CCF	Chief Conservator of Forests
CI	community institution
CPIM	Communist Party of India, Marxist
DFO	district forest officer
FPC	forest protection committee
HFD	Haryana Forest Department
HRMS	Hill Resource Management Societies
IBRAD	Institute for Biosocial Research and Development
IGF	Inspector General of Forests
JFM	joint forest management
MC	management committee
MLA	members of the legislative assembly
MOEF	Ministry of Environment and Forests
MTD	Manakpur Thakur Das (a village name)
NGO	non-governmental organization
NTFP	non-timber forest product
PCCF	Principal Chief Conservator of Forests
PRA	participatory rural appraisal
RO	range officer
SIDA	Swedish International Development Authority
WBFD	West Bengal Forest Department

The loggers girdled the largest trees with black paint strokes. For ten days the trucks hauled away the con-·demned trees from the mixed deciduous forests of village Sagwada in Udaipur District, Rajasthan. A contractor from 'a neighbouring town had won the bid to the harvesting rights for the tract of state forest land. At first, the native inhabitants of these once densely forested Aravalli hills could only witness in dismay the cutting of their natural life-support system. Their alarm escalating, the villagers organized to protest the felling, attracting some 3,000 people from Bhil tribal communities in the surrounding area. In a bold act of leadership, the village Sarpanch (headman) mounted the bus and journeyed to the state capital of Jaipur, camping out at the office of the Secretary of Forests for one week while waiting patiently to plead for a temporary injunction to stay the felling. When the village leader was finally allowed to address the Secretary, his simple statement about the forest captured the tribal's perception of what was happening: 'Our goddess is leaving us.'

— A.J. Mehta, Director
Sewa Mandir, Udaipur, Rajasthan, 1991

Introduction

Mark Poffenberger ■ Betsy McGean

Throughout India, the destruction of natural forests for timber, cropland, fuelwood, pasture, urbanization, and commercial industry has had a profound impact on the lives of millions of rural communities. The deterioration of the nation's expansive forests has exposed critical watersheds, accelerated topsoil erosion and sedimentation of rivers and reservoirs, exacerbated flooding, and overtaxed the land's natural resilience and capacity to regenerate and sustain its productive functions. Attempts to tighten bureaucratic controls over state forests have often led to heightened conflicts among users and further assaults on the ecosystem, rather than conservation and sustainable use.

Yet this is not India's exclusive story. In fact, it mirrors a larger history of the unravelling of much of the global forest estate. The struggle for forest resource control among politicians, private business people, bureaucrats, and local communities is a persistent, all-too-common theme in many developing and developed countries. Newspaper exposés that describe how bureaucrats under political pressure awarded logging rights to contractors and multinational companies at heavily subsidized prices are as prevalent in the United States as in India or Indonesia. Across nations and borders, similarities among the intricate issues of forest management are no coincidence. The problems are deeply rooted in the historical processes through which state forestry institutions evolved over the last century. They reflect concepts of bureaucratic centralization in resource governance, authoritative legislative strategies, and management attitudes and practices that have been borrowed from the temperate industrialized world and widely adopted in many less developed tropical countries. The current global crisis in forest management is profoundly embedded in the past. It will certainly endure far into the future unless societies and their institutions understand the lessons of the past and act upon them.

This book speaks about forest management in India: its past, present, and tentative but promising, future. It documents tales of struggle, conflict, powei, and control. Pressures exerted on the dwindling forests of the subcontinent have mounted exponentially in recent decades. Nowhere are demands on forest ecosystems greater than in India, where human dependencies are staggeringly high and growing rapidly, fuelled by vast human populations, livestock, and industrial demands. With a population approaching one billion and less than 10 per cent of the total land area covered by good quality, productive forest, the human-to-forest ratio is one of the lowest in the world. Yet millions of poor rural Indian families are economically dependent on forest resources for fuel, fodder, food, medicines, housing, cottage industries, and other basic needs. Overwhelming human and livestock pressures on a shrinking resource base would lead many to predict an inexorable decline of forests that ultimately concludes with exhaustion and extinction.

But remarkably, in many parts of India today, poor rural communities are demonstrating that this scenario is not an inevitable outcome. Instead, through collective action, small village groups have begun rallying to protect and reclaim degraded forest lands, banning grazing and logging, and controlling fires. Thosands of villages have organized themselves to reassert their authority over forest tracts in an attempt to reverse degradation and restore productivity. Often with little or no assistance from the state or other outside actors, these protection groups have established access and use controls to facilitate the natural processes of ecological succession and recovery. The key to these reversals seems to lie in the transition from an open access management 'vacuum' to controlled access and monitored utilization. This is achieved through user group-centred controls. As tenurial rights and delineated responsibilities become vested in the user group, conflicts are reduced, communications improve, and local knowledge once again informs decision making. In less than one decade, large tracts of state lands that recently existed as scrub-covered wastes are now regenerating into biologically diverse, closed canopy secondary forests that produce a broad range of forest goods and ecological services.

It is striking how little information actually exists about the role of India's communities in managing the nation's precious forests. Government data — whether forest department working-plan documents, stocking-level figures, budget accounts, industrial-timber requirements, or staff records — fill departmental offices across the country. Yet information about the location and daily forest practices of India's villages is scant or non-existent. Only with the increasing recognition of the role of communities in forest protection under the joint forest management initiative have foresters, non-governmental organizations (NGOs) and researchers begun to collect data and document the growth of the significant grassroots forest protection movement. Outsiders who visit villages in forested districts are surprised to discover both the strength and numbers of communities actively engaged in implementing resource protection and use controls. Frequently located in remote regions where city planners and senior professionals rarely venture, these groups have gone largely unnoticed during their decades-long struggle to reclaim enough management control to protect their forest and water resources.

The response to community resource management initiatives has been mixed. Many foresters intimately familiar with the field realities of policing India's vast public forest lands, with a handful of guards against millions of impoverished people, understand its ultimate futility. Guards and forest rangers are frequently confronted with unrealistic implementation policies and must adapt their own informal, more practical agreements that will better ensure the survival of forests and communities. Certain innovative mid-level and senior officers have been experimenting with various forms of community-agency collaborative management models for years. Yet, until recently, there were few policies or programmes to support or sustain these endeavours. While appreciated by community groups and field NGOs, these individuals have also been criticized and held suspect by other colleagues in the department. Sometimes, more traditional foresters have felt threatened, fearing that the empowerment of communities would erode the authority and legitimacy of the forest department and lead to the chaotic misuse of resources. In fact, field reports from many Indian states indicate that, in most degraded forest areas, forest departments have already lost much of their control over daily access and exploitation. In some forest ranges, field staff risk their lives in valiant attempts to stave off or reduce the illegal flow of timber and fuelwood that are steadily extracted from forest areas. Such uncontrolled access, combined with the absence of incentives among users to monitor overuse, has led to unsustainable cutting and grazing practices and has been a driving force in forest destruction. Struggling to meet their biomass survival needs, forest communities subjected to constant confrontation with the forest department no longer respect the responsibilities of the forest officers, and have been known to literally drive them out of their assigned territories.

The establishment of joint management agreements allows department field staff to redefine their relationship with the community and eventually to regain the trust and alliance of villagers. Only then can the common goal of stabilizing forest ecosystems through co-operative effort be accomplished. The choice is stark: either all parties lose when the forest is destroyed or all benefit through its regeneration and sustainable management. In this latter win-win scenario, the empowerment of community management groups to take the lead reaffirms the forest department's role as state 'custodian,' overseeing and endorsing the work of local forest-user communities.

Some government foresters continue to argue that the provision of usufruct rights to community groups gives too much away. Certain NGO leaders counter that anything less than giving *all* forest benefits to communities is a compromise unacceptable to the forest department. Striking a compromise may be the only politically practical way to resolve this dispute. In the name of India's commitment to social justice, clearly those with the greatest needs and rights to forest resources are the communities and tribal populations who have depended upon them, and lived in the forests for centuries. In most states today, the rights now being offered under recent joint forest management (JFM) government resolutions actually offer little more, and sometimes even less, than what these villages enjoyed under earlier settlement acts and *nistar* agreements. JFM agreements, however, are helping to forge new relationships between rural communities and the forest department. By formalizing and further legitimizing prior or existing rights, they provide the framework for an essential *psychological security* heretofore unknown, enabling communities to invest their labour and time in patrolling, protecting, and managing the forest. Admittedly, poor communities will only have an economic incentive to protect these resources if the state guarantees it will not allow an outside contractor to log the area under community protection once the trees have matured. Further, communities that close areas to grazing and fuelwood extraction forego essential income in cash and in kind while waiting for the forests to regenerate. If they fail in their protection and management duties, the forests will not regenerate, and deferred benefits will also be lost. Hence, a powerful motivation for communities to succeed is intrinsic to the strategy.

JFM programmes should not be viewed as a subsidy or give-away scheme for rural communities, but rather as long-term agreements between rural groups and government, which secure community rights to forest products in exchange for serious responsibilities. In most cases, after a set period of time under voluntary community protection, current joint management agreements are guaranteeing a certain share of the timber harvest to communities, as well as rights to non-timber forest products. More important, these agreements are officially acknowledging the vital *role* for the community in management. In sum, it is the covenant of trust and the co-operative sharing of rights and responsibilities that make JFM a partnership between forest departments and community groups. Given the change in attitudes, policies, and operations implied by JFM, forest departments have little more to offer communities in the early phases than this formal recognition of willing collaboration. Yet, legitimizing the forest protection and management activities of India's rural villages is a critical function. Over time, forest departments will enhance their own capacity to provide more relevant technical information regarding forest ecosystem management, non-timber forest product enrichment, processing and marketing, and local institutional management development.

This book poses many questions regarding India's experiences with community forestry. Is the widespread community response being documented in India today a desperate reaction to resource scarcities and environmental decline? Is it in part a sign of expanding democratization and local empowerment? Why is a grassroots environmental and socio-political movement expanding in some of the poorest regions of India, which also suffer the highest illiteracy rates and the most serious forest degradation? How can local community institutions sustain their systems of forest protection and management once ecosystems recover productivity and become increasingly valuable? What lessons can we learn and transfer from these emerging experiences that might assist India and the rest of the world in managing their natural forests more sustainably as we enter the twenty-first century? Framed in time and space, each chapter attempts to illuminate a piece of this complex puzzle.

Divided into three parts, the book examines the resurgence of community forest management in India and its social, political, institutional, economic, and technical consequences. Part I reviews the historical background and the current status of grassroots forest management initiatives nationwide, and assesses the policies being enacted to support local environmental protection initiatives. In Chapter 1, the authors chart the evolution of rural environmental movements, mapping patterns of socio-political organization and exploring the forces that drive these processes. The very direct geographical overlap between forest regions, areas with severe poverty, and tribal concentrations underscores the importance of poverty alleviation and social equity in forest management. The chapter proceeds to document how communities in diverse regions of India are struggling to protect disappearing natural forests.

Chapter 2 examines historical relations between the state, forest lands, and rural communities. Centralized governmental control over natural forests emerges as a comparatively recent phenomenon of the past two centuries. Since the establishment of British colonial control in the mid-nineteenth century, state custodial authority over forests has been systematically extended. At the same time, the authors show that both as a matter of policy and in the villages, individuals have historically resisted the erosion of community forest rights. Over the past few years, national and state-level policies that support the rights and needs of rural communities to forest resources have been formulated, beginning a reversal of century-old trends. This chapter explores how these policies might help or hinder spontaneous, indigenous management initiatives to regenerate and sustain the forests.

Chapters 3 through 5 further consider how the growth of state control over forest lands has affected rural communities. These document the many conflicting forces that have shaped past and emerging policies. A number of bewildering policies — emanating from the revenue department to encourage agricultural expansion and an expanded tax base, from the forest department to consolidate territorial authority and commercial timber exploitation, from political powers, and from industrial interests — have interfaced and frequently clashed with India's diverse rural communities and their cultural traditions. As these chapters illustrate, the outcome has been conflict and unsustainable forest use. The search to develop sustainable systems of forest management, which respond to national and local needs, remains a continuing challenge. The authors believe that examining the experiences of the past century may help in a socially equitable and renewable forest utilization in the future.

Chapter 3 reviews the position of India's first Inspector General of Forests, Dietrich Brandis. In the mid-nineteenth century, Brandis was one of the first advocates for maintaining the role of Indian villages in forest management. While his unconventional views were not adopted as a policy, Ramachandra Guha emphasizes the striking parallels between Brandis's early vision and managerial concepts, and the evolving, contemporary joint forest management strategies.

Chapter 4 examines the history of conflict between Bhil and other tribal communities in the Dangs of south-eastern Gujarat, tracing the actions of the forest department as it gradually expanded its control over forests. At the centre of the dispute was the state's rejection of traditional communal land use rights and the native agro-ecological practices of local communities. Governmental interest in controlling forest lands and preserving the industrial values of standing timber generated a direct conflict with tribal populations which viewed the forest as their ancestral home. David Hardiman carefully describes the government's opposition to long-rotation shifting agriculture (*kumri*), and its systematic but unsuccessful attempts to force communities to adopt sedentary agriculture. Through revolts and sabotage in the form of forest burning, alienated tribals resisted state attempts to usurp the forest domain throughout the twentieth century. Antagonism between tribal communities and the state continues even today in the Dangs, and many institutions, such as the forest labourers' co-operatives, manipulate the situation for their own political gain. Hardiman concludes that only through the empowerment of small community groups as forest managers can the patterns of 'tribal defeat and loss of land and livelihood' be reversed to stem further decimation of the forests.

Chapter 5 chronicles the saga of tribal struggle in the 'Jungle Mahals', an ancient tribal forest region of south-western Bengal. The tale begins in the mid-eighteenth century, when the diverse hunting, gathering, and agricultural communities of the region were wrestling to maintain their independence from local rulers and powerful landlords (*zamindars*). Their response strategy was one of retreat, dependent upon their ability to withdraw ever more deeply into the forests when encountering outside forces. With the growth of state authority, the alienation of their agricultural lands, and the shrinkage of forest cover over the past 200 years, disempowered tribal groups became increasingly impoverished and progressively lost their ability to physically flee exploitation imposed by both the private and public sectors. Nonetheless, many of these indigenous communities have continued to resist external attempts to usurp their farmlands and erode their rights to forest lands which form the core of their socio-cultural and economic systems. In recent decades, the extended tenure of a populist Marxist government in West Bengal has facilitated the resurgence of community forest management, which has received growing support from the state's forest department and NGOs. The chapter contends that indigenous environmental movements, such as those in West Bengal, provide a sound basis for the broad-based decentralization of public land management. At the same time, these nascent initiatives, and the communities that spawn them, face similar types of political opposition as their ancestors. The challenge is to find ways to strengthen their local capacity to act effectively as protectors and managers of the forest estate.

India's forest communities have faced increasing marginalization for several centuries. While forest departments have grown in size and numbers, their financial and human resources remain woefully inadequate to ensure proper management of nearly one-quarter of the Indian subcontinent classified as public forest land. In Part I, the book advocates that, if given a conducive environment, forest communities can quickly re-establish control over the use of forest and grassland resources, thereby reversing a long process of alienation and disempowerment. This will be far from simple. Communities possess a complex history of prior rights, both formal and informal. Forest departments are firmly entrenched in institutional procedures and regulations and attitudes on both sides are often hostile towards each other.

Part II further explores the role that people and institutions can play in forging new joint management partnership between communities and forest departments. Chapter 6 analyses how India's diverse rural communities and their local institutions can work effectively, illuminating both common limitations and unrecognized strengths. Madhu Sarin describes the processes and obstacles involved in linking people with forests, defining community group membership in consideration of dependency needs, rights, and traditions. The chapter discusses the capabilities that community institutions must develop to act as effective managers, including controlling forest access and use, settling disputes, distributing produce equitably, and interacting positively with the state. Ultimately, forest departments which are interested in helping local communities build institutional management capacity must approach the task in a supportive, rather than directive, top-down manner.

In Chapter 7, Subhabrata Palit examines the challenges facing Indian forest departments as they attempt to build new skills and capacities to work closely with rural communities in microplanning and ecosystem management. While two decades of foreign-assisted social forestry programmes in India began opening communication channels between foresters and villagers, the author notes that these failed to address basic conflicts regarding tenurial rights, equity, and responsibilities over natural forest systems. Although private and community plantation projects were heavily funded, little support was available to improve communal management on public forest lands. Inefficiency, cynicism, and corruption through infusions of foreign finance led to growing confrontations between communities and the forest department, and in some cases even divided the agency internally. The emergence of JFM provides an alternative foundation for collaboration. However, Palit insists that forest departments will need to make fundamental shifts in their attitudes, policies, and procedures to respond effectively to the needs of local forest management groups. Merits such as creative problem-solving, honesty, and field-based achievements must drive the incentive system, rather than seniority or politics. Foresters need to increase their capacity to undertake applied research programmes, provide technical guidance, resolve disputes, and generally support community forestry programmes. Planning systems must become transparent to allow local input into priority setting, while de-emphasizing commercial timber production to respond to a wider range of important non-timber forest products. Finally, forest departments must develop the political will and unity to succeed in devolving authority and respect to community management groups.

In Chapter 8, McGean, Roy, and Chatterjee explore strategies to create awareness, skills, and gender sensitization through working groups, training programmes, and attitudinal changes, which can facilitate co-operative action. Case studies that describe interviews with forest department beat officers and the induction of women field staff into the forest department draw upon experiences from JFM implementation and training programmes in different parts of India. Different strategies and methodologies employed by the NGOs are explored, as are mechanisms other than formal training that contribute positively to the process of change. The authors contend that interactive training with participants can indeed break through attitudinal barriers, but the role of reorientation must be taken more seriously in terms of enhancing India's current limited capacity.

In Part III, Chapters 9 and 10 review fundamental shifts in the way forest management objectives and operations are viewed. These chapters assume that natural forest management will focus increasingly on local economic and environmental needs, with commercial timber production shifting to private lands. They explore the ways in which management processes will need to be intensified to reverse degradation and re-establish healthy and productive ecosystems.

In Chapter 9, Poffenberger reviews the multiple perspectives that can be used to place values on the forest. Projecting into the future, the author explores how current forest management objectives will need to change to meet national, regional, and local needs. Because millions of Indians depend heavily on forest resources for subsistence, their needs will have to assume precedence over those of commercial users. Millions of rural people employed in small, forest-based cottage industries require improved production, processing, and marketing support. At the same time, the inevitable, long-term environmental imperatives of the subcontinent demand a more holistic approach to management. The chapter argues that this will require a serious commitment by planners, donor agencies, and forest department staff to well-targeted applied research and extension programmes. NGOs and university researchers can contribute substantially to the effort if they build their own capacity to respond to sector-specific issues. The author concludes that the quality of life for most rural Indians could be improved by converting wasted areas of former natural forest into intensively managed, biodiverse ecosystems (i.e. through regeneration and enrichment planting).

Chapter 10 examines the implications of shifting management objectives for India's natural forests from timber production towards a more holistic approach of managing the forest ecosystem. This implies de-emphasizing timber plantation and emphasizing natural forest regeneration. The regenerative powers of nature have long been recognized, but only recently has attention focused on reliance upon community-based access controls to stimulate forest regrowth.

The concluding chapter documents the impressive rates of forest regeneration that are possible, especially through coppice shoot rejuvenation from highly degraded forests, if these are protected from further exploitation. The authors note that in many disturbed forest systems, biomass, biodiversity, and the flow of valuable non-timber forest products increase dramatically, even after short periods of three to ten years. In contrast to monoculture plantations, natural forest regeneration requires minimal capital and technical inputs, but far outperforms plantations in terms of many environmental and economic functions. Furthermore, natural forest regeneration has equity implications, often skewing benefits disproportionately to the poorest, including tribal and landless women, whose opportunity costs of collecting forest products are the lowest. For more than a century, India's natural forests have primarily been managed to optimize the production of commercially valuable timber. As community-based user groups become increasingly prominent in management decisions, forest management objectives will shift towards multiple product needs. The chapter explores various approaches to restructure forest management strategies, both in terms of goals and operational techniques, to respond to varying environmental and economic objectives.

In conclusion, the authors attempt to synthesize historical experience, case materials, and policy analysis to understand how forest conflicts and their solutions are linked temporally and spatially. India is one of the first nations to seriously consider policies and programmes that decentralize and 'de-bureaucratize' management of its public forest domain. These historic changes appear to be driven by environmental crises affecting some of the nation's poorest rural people, by the growth of democratic institutions, and by the commitment of certain leading policy-makers, scientists, and foresters to social justice and environmental restoration. At the same time, bureaucratic conservatism, scepticism, and entrenched vested interests inevitably impede such changes. The debate of the past two centuries over annexation versus populism continues today; only the stakes have risen as the importance of forests to a growing, biomass-dependent population expands.

The resurgence of community forest management on the subcontinent offers significant lessons to many developing nations engaged in similar struggles to stabilize and sustain their forests while alleviating rural poverty. In the vanguard of a grassroots environmental movement which is spelling public land reform, India is a laboratory of experimentation and learning. The inevitable change may have arisen first from the extreme pressures on natural forest ecosystems and the consequent degradation they have suffered, both from commercial and local exploitation. But the heavy dependence of India's massive tribal and other rural populations on forest resources have left them no choice but to reassert their control. Nature has responded with vigour to the care that local villagers are providing, rewarding some of the planet's most dispossessed and marginalized populations with steady flows of biomass products upon which they survive.

The rethinking of forest management goals in India demands new strategies that are community centred and place priority on satisfying local socio-economic and environmental needs. Moving far beyond the archaic model of centralized, custodial state management, through the various trials and phases of social forestry, the enlightened progression emerging in India today is defining a new paradigm of 'ethnoforestry'. The rationale involves ecosystem dynamics, indigenous community knowledge and creativity, local empowerment, faith in community self-reliance, and full management participation by those people who are closest to the forests, strongly dependent upon them, and highly motivated to keep them productive. After all, the largest group of environmental managers in the world are farmers, forest farmers, and forest dwellers — more than half of whom are the planet's rural women, the key actors in biomass-based economies.

The authors of this book unite in their common belief that the seeds of India's future forests lie in indigenous systems of management, the unleashing of creative human resources, and the assurance of tenurial incentives for forest-based communities. India's grassroots experiences of ethnoforestry can be replicated across regions and nation-states, provided communities are allowed to regain management authority over the integral resources that sustain their livelihoods. To this end, political powers and influential donor agencies will need to restrain commercial and financial interests, and instead encourage forest bureaucracies to facilitate the socio-political processes of public land reform.

Part I

The Resurgence of Community Resource Management: Control, Conflict, and Compromise in the Forest

Chapter 1

Communities Sustaining India's Forests in the Twenty-first Century

Mark Poffenberger ■ Betsy McGean ■ Arvind Khare

Understanding the Problem

The forces driving deforestation in India are complex. There is an urgent need to understand the relationships between resource degradation and social unrest which threaten both the environment as well as the nation's social and economic stability. India's forest ecosystems have suffered from extensive, successive disturbances over the past century. Today their existence is threatened.

Statistics concerning rates of deforestation are misleading. They imply that forests are either present or gone. Although some forests do disappear abruptly through clear-felling or devastating fires, most forest ecosystems instead suffer a *process* of degradation. This occurs through a series of human interventions that result from a lack of management controls. Often, responsibility for deforestation is cast on a single user group, be it loggers, swidden farmers, or women fuelwood headloaders; yet, more often, multiple actors are involved in disturbing the same tract of forest at different points in time. Millions of hectares of forest land are overexploited through selective logging, illegal cutting, grazing, migrant farming, and fire. Most of India's forests are degrading over time, 'ratcheting down' biologically as they lose biomass, diversity, and topsoil, eroding their complex structural and functional integrity.

Attempted solutions to the problems of deforestation are often misguided and ineffective. Too commonly, they are defined in terms of capital investments, state-of-the-art technologies, and enhancement of modern professional capacities. Yet, the huge foreign investments and new technological strategies of the past decades have had relatively little impact, as witnessed by unabated rates of forest degradation. A recent World Bank report noted, after spending $ 1.5 billion on forestry projects in Asia between 1979 and 1990, 'The Bank's investments have had a negligible impact on borrowers' forestry sectors as a whole.'[1] Even in the well-funded, best-protected 'Project Tiger' parks in India, the amount of forest land classified as degraded increased by an estimated 186 per cent between 1983 and 1989, while good quality forest, with canopy closure of more than 40 per cent, declined by 50 per cent during the same period.[2] If the most intensively 'managed' and heavily-funded wildlife parks in India are deteriorating at such a relentless pace, reserve and protected forests with far fewer guards and much smaller budgets appear to have an even lower chance of surviving.

Forest management systems evolving since the nineteenth century colonial era are still largely premised on models of unilateral, centralized state control. Public forest lands cover 25 to 75 per cent of Asia's total land area. Currently, most South and South East Asian governments still possess sole legal rights to virtually all of their natural forests. In India, an estimated 97 per cent of all forests are owned by the state.[3] The forest agencies entrusted with the protection of these lands comprise mainly bureaucratic staff, most of whom are office-bound and heavily burdened with administrative duties. Despite their limited field time, they are responsible for monitoring the forest use of millions of rural inhabitants and migrants, as well as loggers and livestock. The ongoing failure to stem forest degradation Asia-wide indicates that forest departments alone are simply incapable of such an unrealistic mandate. With the rapid expansion of human populations and the transformation of national politics and economies, the world has changed dramatically. Rural communities in India have growing political power to demand rights to manage the local forest resources upon which they depend. Elected political representatives are gaining influence under emerging local-governance systems (*Panchayati Raj*) and are attempting to respond to the concerns of village constituencies. India's natural forests have suffered severe degradation over the past two centuries. Deforestation has accelerated even faster during the last three decades. Remote sensing of India's forest cover between 1980 and 1987 indicated that deforestation was proceeding at a rate of 1.48 million hectares per year, far more rapidly than previously believed.[4] More recent imagery for 1989–91 suggests that deforestation may be slowing, possibly due to the immense national investment in fast-growing tree plantations.[5] Yet, plantations of fast-growing species cannot be equated with natural forests. Monoculture eucalyptus and other plantation species do not possess the biological characteristics of natural forests. Further, plantations are not generally established for environmental reasons and are usually felled within a decade of establishment. By the early 1990s, estimates of good forest cover in India, with a crown density of at least 40 per cent, ranged between 9 and 13 per cent of the land area.[6] These figures stand in stark contrast to the national goal of maintaining one-third of the nation under forests.

It has been estimated that India's forest-dependent populations minimally require 0.5 hectare of forest land per capita, while the mean availability is only 0.1 hectare or one-fifth of that required.[7] Further, India's burgeoning population, which is now approaching one billion and continuing to expand rapidly in poor rural areas, faces a shrinking pool of forest resources. Given these trends, a predictable response has been their continuous overexploitation and further degradation.

Formal economic indicators fail to reveal the importance of forest resources for subsistence and informal users. In the 1970s, decisions to make major capital investments in industrial forest development in Bastar and other areas were based on the forest sector's minor contribution of only 1.3 per cent to the GDP and its ability to formally involve only 0.2 per cent of the national industrial labour force.[8] Yet, India's forests are critically important in meeting a wide diversity of needs for millions. Natural forests directly contribute to the survival of more than fifty million of the world's poorest tribal people. Millions of rural people depend heavily on informal sector forest-based livelihoods.[9] In addition, millions of lowland farmers rely on upland forests and the watersheds they protect to control flooding and to provide a stable supply of water for irrigation. India's vast population of city dwellers also depend on water and electrical power originating far upstream.

A major objective of modern forest management has been to optimize the production of a few valuable timber species. Until the 1970s, foresters relied primarily on manipulating natural forest ecosystem processes to enhance timber productivity. Plantations of exotic timber species were established because degraded forests were failing to regenerate naturally. Natural regeneration was frequently slowed down by ongoing disturbances within the ecosystem. Forest agencies and concessionaires were unable to adequately protect forests after initial logging. At the same time, overexploitation eroded the natural resilience of the ecosystem to generate new growth. Traditional community forest-use practices and protection systems also broke down as indigenous rights declined, and local authority to control access and protect forest lands was lost. Simultaneously, growing local and migrant population pressures often reduced fallow periods, diminishing soil fertility and pushing rural farmers onto more marginal forest lands. Without effective protection to prevent further exploitation and allow for a period of recovery, the natural process of secondary forest succession is slowed or ceases entirely, with a further occurrence of ecological degeneration.

As poorly protected natural forests repeatedly failed to regenerate, and as industrial demands for raw materials rose, foreign donors and national governments began advocating the adoption of fast-growing plantations and increased capital investments. Ironically, the lack of access controls, which were driving the degradation of natural forests, also generated similar management problems for plantations. In both cases, effective protection was a fundamental prerequisite for forest stability. In fact, monoculture plantations of exotic species often proved to be even more vulnerable to natural disturbances and far more costly to establish. Driven by growing concerns over timber requirements and rural fuel supplies, the Indian government, often assisted by donor funds, initiated a massive social forestry programme. Billions of fast-growing trees, primarily eucalyptus, were planted in India. The strategy sought to raise wood for local needs on common and private land, taking local pressures off natural forests so that they could be used for industrial purposes and environmental conservation.

Social forestry, and particularly farm forestry, was successful in increasing the availability of construction poles, pulpwood, and small timber. The programme also began to sensitize forest department staff to community needs. Yet, the broader social forestry programme failed in a number of areas. It did not relieve pressures on natural forests, which continued to degrade. In fact, it can be argued that in many states the reserve and protected forests received less attention from the forest department, since the staff became heavily occupied with the establishment and administration of social forestry plantation projects. Social forestry programmes were often not designed to respond to local community interests or institutional capacities. Communities were frequently dissatisfied with placing responsibility for plantation management with politicized panchayat organizations. In many communities, members felt they would never receive any tangible benefits from the project, and very often they did not. Finally, and most critically, social forestry programmes failed to respond to or resolve conflicts between communities and government over rights to natural forest lands. This failure led to their continued overuse and has frustrated all attempts to stabilize forest use and allow ecological regeneration.

Around 1985, there was still a sense among senior Indian planners that plantations could continue to help achieve the broader goal of restoring forests on an estimated 70 million hectares of degraded land.[10] Yet, experience indicates that the nation was only able to replant 3.7 million hectares between 1950 and 1980, with little information regarding the survival of these plantations, while the pace of deforestation during the 1970s and 1980s was averaging 1 to 1.5 million hectares annually. Even the targets of the Sixth Five-Year Plan anticipated planting only 0.4 million hectares per year.[11] While B.B. Vohra states, 'The future of our forestry will therefore ultimately depend on the rate at which we can bring denuded areas under fresh plantations . . . ,' he also acknowledges that this will depend on protecting them until they reach maturity. He notes, 'The very serious limitations from which our forestry establishments suffer in this field arise mainly from their traditional lack of rapport with local populations.'[12]

The sheer logistical problems involved in covering millions of hectares appear far beyond existing institutional capacities. The monetary investments required are also staggering. As Vohra acknowledges, 'as every insider knows, the generous subsidies and soft loans associated with many schemes . . . offer an excellent opportunity for corrupt elements to siphon off a portion of government . . . funds.'[13] In fact, large investments may tend to further corrupt, rather than build, institutional capacity. Finally, covering India's degraded lands with fast-growing, short-rotation monoculture plantations, from an ecological and economic standpoint, may not respond to the nation's environmental needs or to the requirements of her rural population.

In recent decades, some communities have begun protecting natural forests either on their own initiative or with the encouragement of forest department staff. Rather than rely on exotic species, communities attempt to regenerate the native ecosystem by protecting it from grazing, fires, and cutting. Recently referred to as joint or participatory forest management, this approach to forestry is very distinctive when contrasted with the social forestry programmes of the past decades. Some basic differences in the goals and operational strategies of conventional social forestry programmes and emerging joint forest management initiatives are outlined in Table 1.1.

India is faced with millions of hectares of forests that are either badly degraded, or are in the process of being degraded. Unless effective access and use controls are established, these ecosystems will continue to lose biomass, biodiversity, topsoil, and the ecological resilience to recover naturally. No capital investments or technologies can respond to the lack of access controls or to the scale of the problem in the time period required. Only through the resolution of conflicts at the local level can sustainable management systems be established. Increasingly, natural regeneration is recognized as the only practical solution to restore much of the nation's degrading forest resources. Community protection of these degraded forests is increasingly viewed as the key element to facilitate this process of ecological restoration. There is a growing movement worldwide to move away from industrial forestry based on plantation technologies towards a holistic or new forestry that emphasizes management of the entire ecosystem, including soils, herbs, shrubs, as well as tree species. This approach can rely both on indigenous wisdom gathered over generations of experience as well as modern science.

People Creating Solutions

A growing number of foresters and planners acknowledge that one of the most promising strategies to stabilize forest resources may be through creating partnerships between rural people and forest agencies. When given clear rights and responsibilities, disempowered forest communities are proving they can work as allies with government field staff and non-governmental organizations (NGOs) to establish effective access controls and install regulated forest-use systems. This strategy, however, implies a massive transfer of responsibility to thousands of forest communities throughout Asia. A strong political commitment from the government is essential if it expects to successfully devolve and delegate authority to rural communities. This transfer will often require a shift away from powerful private commercial interests at the risk of alienating them. Many forest departments in Asia are just beginning to understand the advantages of working with forest communities to build coalitions that can protect and rehabilitate forests. Yet, government agencies possess limited experience with co-operative endeavours involving rural people as equal decision-makers and partners. Past alliances of foresters have primarily been limited to industry. Techniques, procedures, and institutional norms for decentralizing forest management have not yet been well developed or tested.

There is a growing discussion regarding the advisability of acquiring large loans for technically oriented plantation projects. Instead, planners are beginning to recognize the need to address more fundamental institutional and political problems that drive forest destruction. While the process of change urgently needs to be accelerated, Asian foresters, NGOs, and social scientists are beginning to work collaboratively to adapt their strategies to meet a changing forest management environment in the next century.

The strength behind these initiatives is not in the urban centres but is emerging from the villages. India appears to be leading the way. Many communities in India are taking action to protect their threatened natural forests. Tribal communities, particularly, are building upon traditional resource management practices, as well as developing new strategies to gain authority over forest lands and water. Not surprisingly, these grassroots environmental efforts are most common in the poorest regions, where villagers are suffering most from growing resource scarcities. Many of India's tribal people depend on forest tubers as a staple food for six months of the dry season each year. If the forests cannot be stabilized to meet such compelling subsistence needs, their disappearance may eventually dismantle entire villages, forcing villagers to migrate to urban slums and in the process destroy their economy, community, and traditions. Instead, they will place an immense burden on urban infrastructures and create growing dependencies upon the state for their survival.

Fortunately, in response to resource scarcities and poverty, as many as 10,000 communities may now be involved in overseeing the forests in eastern India. Relying on local leaders, village councils, volunteer patrols, indigenous knowledge, and consensual decision-making, a growing number of communities are stabilizing and regenerating their forests. This experience stands in stark contrast with conventional development strategies that depend on external capital and new technologies managed by government officials. It is a humbling thought to recognize that rural people are accomplishing what the investment of millions of dollars could not. This chapter explores the resurgence of grassroots environmental protection and management in India and attempts to summarize the historical, human-ecological, and political contexts from which these grassroots movements have arisen.

Historical Patterns of Community Forest Protection

While India's rural environmental movements are poorly understood and documented, they appear to be gaining momentum and receiving support at the village, state, and national levels. Local environmental activism is not a new phenomenon but is rooted in the past, with strongholds in certain regions of the country. One study identified sixty-four incidences of major tribal revolts between 1778 and 1971, most of which reflected repeated uprisings by the tribals of the Chotanagpur Plateau, the Bhils of Gujarat, and the tribes of north-eastern India.[14] Often, resistance was triggered off by encroachment on agricultural land, though there is increasing evidence that tribal concerns over the loss of their forest resources was an important factor in many insurrections.[15] The curtailment of tribal rights and privileges over forest resources under the 1894 national policy on forests was a major blow to the rights of forest dwellers, initiating a process of alienation from critical resources. For more than a century, tribal people have been squeezed between the alienation of their land to moneylenders and high-caste Hindus and the loss of their forest resources to contractors and state agencies. These have led to a deep-seated antagonism between tribals and foresters. As B.B. Vohra notes:

Foresters have always looked upon the villager and his animals as a big nuisance and would ideally like to be given full powers to keep them out of even 'protected' and 'unclassed forests' just as they do in 'reserved forests.' This attitude has created a big gulf between the forester and the villager and shows how little the former understands the problems of the latter. If only our foresters had made a sincere effort to explore the possibility of placing denuded areas under tree cover with the co-operation and the active participation of local communities, the vicious circle created by growing scarcities of forestry products and . . . tensions between villagers and forest department functionaries, could have been avoided.[16]

Apart from an armed struggle, tribal communities throughout India have sought greater control over their natural resources through political movements. Lobbying for a separate tribal state (Jharkhand) in sixteen districts in eastern India has been under way for more than six decades. Separatist tribal movements have also emerged in the Gond area comprising the old Central Provinces, as well as in eastern Gujarat, southern Rajasthan, and northern Maharashtra. In some cases, tribal communities have allied themselves with communist insurgency groups (Naxalites) to better empower their struggle. In explaining the incidence of violent tribal struggle, one anthropologist notes that a primary factor has been the 'harshness of the forest laws and regulations and the lack of sympathy and understanding in administering them.'[17]

For decades, many political leaders, social scientists, and foresters have championed the cause of India's tribal communities, and much progressive legislation was passed to protect their land tenure rights. Notwithstanding these sincere efforts, the position of tribals has declined with the degradation of their forest resources, the shrinking of common property grazing lands, and the loss of agricultural land. Driven to poverty, tribal communities, which once practised more sustainable forest management, began over-exploiting local resources — including overgrazing pastures and forests, shortening rotation periods on agricultural land, and intensively hacking and uprooting trees to obtain firewood for commercial sale in order to survive. As the resource base degraded further under these pressures, the cycle of poverty, migration, and social erosion has intensified.

Yet, as forest resource shortages became acute, in some areas village leaders have drawn attention to the problem, suggesting strict forest controls. Voluntary patrols to regulate forest use have been successful in many villages, allowing rapid regeneration to take place, while encouraging other communities to adopt similar management systems. The spread of these initiatives is apparent throughout south Bihar, south-west Bengal, and Orissa, and are now being identified in southern Rajasthan, Gujarat, the hills of Uttar Pradesh, and parts of Madhya Pradesh and Maharashtra.

The pace at which this grassroots environment movement has spread in eastern India during the 1980s is unprecedented in Asia. Case reports indicate that communities have taken action in response to forest resource scarcities upon which their survival depends, and fears of micro-climatic and hydrological changes that adversely affect their lives and agricultural systems. Villagers are increasingly cognizant that they cannot rely on the government to resolve their resource crisis, and that only through their independent local actions can they meet their natural resource needs. That community acquisition of state forest lands has not been opposed by state agencies appears to be tied to their diminished value. To the extent that forest departments are passing supportive orders for community forest protection, it is generally limited to badly degraded areas.

During the 1970s and 1980s, while state forest departments were preoccupied with large, plantation-oriented social forestry projects, community forest protection groups began emerging with little attention from forest agencies. With the exception of West Bengal — where a few progressive foresters, encouraged by the state's populist government, actively supported and facilitated the emergence of Forest Protection Committees — no formal programmes or projects were initiated to foster the spread of local forest protection. It was not until the scale of grassroots forest management initiatives began to be recognized in the late 1980s that the national and state governments began to perceive its significance and acknowledge the need to recognize and legitimize community efforts. In 1988 and 1989, Orissa and West Bengal passed state resolutions recognizing the validity of community forest protection. In June 1990, the government of India passed guidelines notifying that exclusive rights to forest products be extended to those villages effectively protecting public forest lands. By 1994, sixteen states passed similar orders. With the support of these resolutions, and under the banner of joint or participatory forest management, a mechanism is evolving to facilitate communication and co-ordination between forest villages and the government.

Although the primary objective of joint forest management is to ensure sustainable use of the nation's forests to meet local needs equitably while achieving India's broader environmental goals, some contend that the forest department's recognition of independent community forest management activities may be used to co-opt them and bring them under government control. In some areas villagers are extremely wary of the forest department's involvement and have banned them from entering their areas. In other regions, however, they are anxious that the forest department staff register their groups and demarcate their protected forests.

Most state forest departments have now been authorized to establish formal dialogues with communities and devolve some management responsibilities to them for the forests that are central to their survival. Still, many conflicts exist between village expectations and needs, and state guidelines for forest management and product sharing. It may be best to view community initiatives to re-establish effective forest management systems as an initial reversal of more than a century of state consolidation of forest control. How effective forest departments will be in establishing productive partnerships with communities remains to be seen. The devolving of management responsibilities for India's forests back to communities will require decades. We can only recognize that it is under way and that it is part of a historic process of social change likely to be irreversible.

Regional Patterns of Local Environmental Activism

Strong associations exist between places where communal resistance to state usurpation of local resource rights has occurred historically and where contemporary initiatives are prevalent. India's existing forests are primarily concentrated in three regions: the Himalayan band to the north; the central forest belt with its nexus in the Chotanagpur Plateau of Orissa, Bihar, and Madhya Pradesh; and the north-south belt of the Western Ghats (see Figure 1.1). Significantly, the locations of India's predominant tribal populations are closely superimposed on the nation's forest tracts (see Figure 1.2). With the greatest economic dependence on forest resources, it is not surprising that tribals possess the most extensive knowledge of India's forests, as well as the strongest motivation to ensure the continuity of these ecosystems. There are also strong correlations between the locations of tribal people, forests, and India's concentrated poverty areas (see Figure 1.3). Central eastern India, where community environmental activism appears most prevalent, is also the region with the greatest poverty, with more than 60 per cent of the population falling below the poverty line. Figure 1.4 reveals the close intersection of forests, tribals, and poverty areas in central India. This chapter examines patterns of community forest protection in three major Indian forest zones including the central tribal belt, the Himalayas, and the Western Ghats. In other parts of India, there are numerous cases of community forest management that remain undocumented and urgently require study and support.

Central India Tribal Belt

One-third of the planet's tribal people live in India, most of them in a broad strip of forests that transects the subcontinent. In 1991, the majority of India's 60 million tribals lived inside or near the *sal*, teak, and acacia forests of the central region, stretching from Bengal to Gujarat. The central Indian plateau is the source of some of the nation's largest river systems. This region plays a crucial role in moderating the climate and regulating the hydrology of much of Peninsular India. By slowing and absorbing run-off, these forests sustain or contribute to the Ganges, Narmada, Bundelkhand, and Tapti rivers. Yet the forests of central India have suffered extensive disturbance over the past century. The push to construct roads and railways from the mid-nineteenth century onwards encouraged contractors to fell and extract huge quantities of high-value teak and *sal* timber, leaving depleted forests and damaged ecosystems in their wake.

For thousands of years, forests have been an integral part of tribal traditions, culture, and economy. Fiercely independent, they have for centuries attempted to preserve their autonomy and lifestyle by resisting outside attempts to exploit their forests. However, over the past two centuries tribal people have been increasingly subjugated by governments, landlords, and moneylenders. The loss of their agricultural lands through privatization, and the degradation of their forests has left them alienated, disempowered and vulnerable to exploitation. As a consequence, the tribals of central India are among the nation's poorest social groups. Due to their isolation and minimal political influence, these communities have had little access to education, health, or other economic assistance. Yet, while many tribal people rely on subsistence-hunting, forest-gathering, and shifting cultivation for their livelihood, they possess a strong sense of cultural identity and have often been able to maintain an economic independence that has kept them united when facing exploitation by more powerful groups.

In both eastern and western India, community-led forest protection and management are often organized around small hamlets, sometimes referred to as *sahi* or *phalia*. Once community groups have re-established local controls, the dominant sal forest ecosystems of the east and the teak forests of the west often experience rapid biological recovery and regeneration. Typically, in the degraded sal forests of eastern India, after five years of community protection, the knee-high scrub and brush forest is transformed into a multi-tiered, 7 to 8 metre closed canopy secondary forest. In one research area in south-west Bengal, after five years, biodiversity increased from several dozen vegetative species to 214, nearly 90 per cent of which were used by the Santhal tribals for a wide range of domestic and commercial purposes.[18]

Until recently, in both Orissa and Bihar which possess large tribal populations, forest departments, government agencies, and even NGOs have played a minor role in catalyzing or supporting such village forest management initiatives. While the West Bengal Forest Department facilitated the formation of Forest Protection Committees (FPCs) through supportive field staff and a sympathetic leadership, it has done so without special budgets or projects. As a few communities began protecting some forests and closing access to others, their actions often encouraged neighbouring villages to take the responsibility of protecting adjacent forest tracts. For example, in Sarangi Range of Dhenkenal District in Orissa, a few villages had protected forests since the 1950s; however, by the early 1980s many communities facing resource scarcities began to establish forest access controls. By the mid-1990s, virtually all of the 20,000 hectares of state forest lands in the territory fell under the protection of individual hamlets.

In many areas, the spread of local forest protection efforts appears to begin gradually, with one or two communities attempting to establish access controls. Once these villages demonstrate that they can successfully restrict entry, and forests begin to regenerate, neighbouring communities adopt other forest tracts until much of the public land is claimed. Sometimes, the authority of the village protection groups is challenged, with neighbouring communities organizing a 'mass loot' where the entire patch of regenerating forest is felled in a day or night. Such incidents, however, are uncommon. More typically, stronger hamlets with effective leadership begin protection activities and are able to sustain them, inspite of initial protests from other villages. The support of local forest department field staff is often an important element in legitimizing forest protection efforts. A regional overview of community activities highlights a similar approach among states towards motivating social and ecological conditions, influences, and local strategies to stabilize forest resource drawdown.

In West Bengal in the early 1970s, the communist government (Communist Party of India Marxist) was building its political base on land reform and other programmes designed to gain popularity among rural communities. After several tribal deaths occurred in Purulia district during a protest over forest policies, the forest department was informed by political leaders that it could no longer use police to repress local forest use. This and similar incidents forced the state government to review its *modus operandi* and to restrict the use of force in resolving resource conflicts.

At the same time, at the Arabari research station in Midnapore district, the programme director was becoming increasingly frustrated with the continuous disturbance of the research site by livestock and fuelwood collectors. After considerable discussions and negotiations, the officer reached an informal agreement with ten surrounding villages to provide them with a share of the timber revenues and all non-timber forest products if they agreed to protect the forest lands. The agreement worked remarkably well;

forest regrowth was rapid and prior tensions and conflicts declined markedly. Over the next decade and through the mid-1980s, while the forest department had no formal programme of support to community management groups, it did not interfere in their local protection activities. During the second half of the 1980s, senior forest officers in the south-western part of the state began to strongly encourage their staff to assist in the formation of village FPCs. The active support of the forest department field staff and informal legitimization further accelerated the establishment of community-based access controls. Between 1985 and 1990, the number of functional FPCs doubled each year, increasing from fewer than 50 to nearly 2000, and covering 320,000 hectares.[19]

The first West Bengal state resolution, which provided official legitimacy to community management groups, however, was not passed until June 1989, about fifteen years after experiments with joint forest management had been started. Since then, both within the forest department and the state government, the growing success of community forest management has received greater recognition. In the early 1990s, in part through a World Bank loan, more systematic support services — including the development of micro-management plans with the community, extension advice, and the availability of small amounts of project funds for forest development activities — were made available. The increasing commitment of the West Bengal Forest Department to decentralized management resulted in a strategic decision to attempt to bring, wherever community interest permitted, all remaining state forest lands under joint management by 1998. To that end, West Bengal has gone further in terms of its political commitment and field programmes for community forest management than any other Indian state or any other nation in Asia.

In the tribal districts of southern Bihar, communities have also become increasingly engaged in forest protection since the early 1970s. Grassroots community forest protection activities are widespread. Many villages had initiated forest protection groups in the 1970s in response to growing forest product scarcities and threats of exploitation by outside groups. The widening influence of the Jharkhand political movement also provided leadership and encouragement to organize around environmental issues. Generally, these community groups received little support from the state forest department. In fact, researchers found the villagers distrustful of the forest department, and in many cases, had banished departmental field staff from their areas upon threat of physical harm. Village committees have attempted to restrict forest department felling operations in their area. In recent years in Hazaribagh West Division in south-eastern Bihar, of the sixty-nine coupes slated for logging, only seven could be harvested since tribal communities burned or threatened to burn logging trucks if they entered the area.[20] As in other parts of eastern India, the degraded sal forests regenerated rapidly through the growth of coppice shoots and associated species.

At the same time, the Bihar Forest Department had been actively engaged through the 1980s in extending a social forestry programme funded by the Swedish International Development Authority (SIDA). Like similar programmes in other states, the establishment of nurseries and fast-growing tree plantations was encouraged. While strategies to form village forest committees and to develop joint management plans were a component of the programme, these activities remained separate and isolated from the indigenous community forest management initiatives, which were expanding throughout the southern part of the state. Under the programme, the forest department established social forestry woodlots covering approximately 5000 hectares per year.[21] By contrast, since 1990, 1242 village groups were registered, protecting more than 600,000 hectares of public forests.[22] Since many village groups remain unregistered, the total number of FPCs is likely to range from two to three thousand. In one survey in Hazaribagh District, thirty-two communities were involved in forest protection, representing 15 per cent of all villages in the area.[23]

It is instructive to analyse why the forest department's social forestry programme was unable to interface more effectively with spontaneous community initiatives. For one, the externally funded social forestry programme was based primarily on models developed during the 1970s for the delivery of financial and technical assistance. This required forest departments to unilaterally determine what would be spent, where, how, and when, with a focus on common and private lands. While in some cases communities may influence the selection of species planted, the authority for the projects rests largely with the agency. In contrast, local communities were attempting to empower themselves to unilaterally facilitate the regeneration of forests and build the institutional authority and management capacity to establish access controls. Despite their efforts, many villages faced serious difficulties in sustaining their new management systems. Frequently, through either internal or external pressures, protection systems broke down and the forests were felled. In one district with sixty-five FPCs, 70 per cent had collapsed. In some areas it was reported that timber merchants, in collusion with corrupt forest department field staff, attempted to undermine community protection efforts in order to gain access to the timber.[24] In other areas, when the value of the forest reached a certain level, often after three to four years of protection, internal community pressures to exploit the resource overcame the commitment to regulate its use. Some communities decided to exploit their regenerating forests for survival, reflecting severe economic hardship. In other cases, neighbouring villages initiated mass looting of a community-protected forest. Yet, despite their collapse, often within a year or two of the felling, many villages attempted to re-establish their forest management systems. Clearly, indigenous community forest management groups in south Bihar are fragile and vulnerable to collapse, and may benefit from external support to enhance their legitimacy and provide economic assistance during severe hardship.

In recent years, certain committed foresters in the Bihar Forest Department have begun seeking new ways to encourage village forest protection groups. Villagers in Bihar have stressed the need to be given unequivocal access and benefit rights to the forests they protect. They have requested the right to protect the forests through their own customary management systems, including the authority to employ local people who will be accountable to the community, as forest guards. They have also suggested that communities without forests should have depots that can supply them with forest products.[25]

Orissa is one of India's most important forest and tribal states. Thirty-eight per cent of the land area is designated state forest, and is inhabited by dozens of tribal groups. Due to the predominance of tribal people and their heavy dependance on forest ecosystems, deforestation has had a devastating impact on community subsistence livelihoods. In June 1993, it was estimated that ten million people were affected by famine.[26] The six worst-hit districts have experienced declining water tables and recurring droughts since 1965, both phenomena perceived to be linked to the deterioration of the region's forests. Commercial logging, followed by small-scale fuelwood headloading and overgrazing, have resulted in widespread deforestation, drastically reducing the flow of forest products which are critical for the survival of many tribal communities.

Fortunately, traditional village-governance systems among both tribal and non-tribal communities have maintained greater operational strength in Orissa than in many other Indian states. Some social scientists believe that because the village panchayats in Orissa were established to encompass larger populations comprising ten to twenty communities, indigenous management systems retained their function, actively fulfilling the needs of single communities.[27] These hamlets (sahi) are frequently comprised of fairly homogenous tribal, clan, or caste groups and have traditionally been involved in a range of community-governance activities, including the management of village water-ponds, temples, and gardens, and the protection of forests and agricultural lands.

In response to the increasing resource pressures and deforestation, many communities in Orissa began protecting local forests in the 1970s. By the late 1980s, an estimated three to four thousand communities had established control over about 10 per cent of the state's reserve, demarcated, and undemarcated forests, covering some 572,000 hectares.[28] On 1 August 1988, the government of Orissa passed the nation's first forest policy resolution endorsing community management. Like many of its successors, the Orissa notification was conventional in its assumption that the government would take the lead in establishing a new 'scheme,' defining new groups and their institutional structures, rather than simply recognizing and supporting existing local forest-protection systems.[29] Furthermore, as a populist political strategy, the chief minister ordered the forest department to form 5000 new FPCs by the end of the same year. During the next six months, district forest officers, range and beat officers, and guards formed and registered several thousand new community groups, while also registering some of the existing groups. The majority of the well-intentioned new groups failed to function while active traditional groups were often unrecognized. Still, by the end of 1993, about

27 per cent of Orissa's state forests were under some type of community control.[30]

The expansion of community management in Orissa — despite difficulties faced on account of the government's overlap interventions and an inadequate policy resolution — is nonetheless significant and promising. Yet the scale of the problem remains daunting and underscores the imperative need for the rapid expansion of community access controls. Between 1983 and 1987, satellite images of Orissa revealed a 10 per cent decline in forest vegetation cover.[31] Reserve forest lands which received more intensive protection by the forest department suffered to a lesser extent, but the less intensively monitored protected and unprotected forests were heavily impacted.

Despite the fact that Madhya Pradesh possesses much of India's best forests and 22 per cent of its tribal people, little systematic study has been undertaken to examine the current trends in grassroots environmentalism and activism in this vast state. It is likely that the expansion of community protection groups in neighbouring south Bihar and western Orissa has also spread across the borders into Madhya Pradesh. Better known in Madhya Pradesh is the work being carried out in the south-central region by communities and forest officers.

The forests of the Hoshangabad area had been subjected to extensive logging, fuelwood cutting, and overgrazing, yet the rootstock remained viable. In 1991 young foresters, influenced by experiences from West Bengal, began encouraging tribal communities to organize protection groups. The district forest officer conducted a series of meetings with his field staff to reorient their approach to closely collaborate with community groups. By 1992, in Hoshangabad District more than 150 FPCs had been formed bringing 75 per cent of all divisional forest lands, totaling 105,000 hectares, under community protection. Communities now patrol the forests on a rotational, voluntary basis, imposing fines on illicit users — including colluding forestry field staff. All grazing has been banned, while fodder grass is cut and carried, raising yields substantially.[32] Grounded in the strong resource dependencies of the tribals and the committed work of local foresters, the rapid growth of community forest management groups in Hoshangabad reflects the possibility of immense opportunities for decentralized management and increased forest productivity for some of India's poorest and most biomass-dependent rural inhabitants.

Farther to the west, the states of Rajasthan, Gujarat, and Maharashtra are characterized by large Bhil tribal communities as well as other smaller tribal groups. Drier deciduous forests are often dominated by teak, terminalia, or acacia species. In Rajasthan, community resistance to forest loss dates back to the eighteenth century. Led by women, a tribal community in the Thar desert hugged the *khejri (Prosopis cineraria)* trees in a high-risk but successful protest to protect them from being cut by a raiding army. Living in a harsh and arid environment, Rajasthani tribals and nomads have historically relied on their extensive ethno-botanical knowledge to sustainably manage their fragile environment. The Bishnoi tribe of the deeper desert, for example, are so named for the twenty-nine rules that guide their society, including environmental conservation and use-rules which are treated as an integral component of their lifestyle.

Though rapid deforestation in the forests of southern Rajasthan since the 1950s has inflicted considerable suffering on local forest communities, villages have responded in different ways to the degradation. In the early 1990s, in southern Dungarpur district, several dozen villages began blockading roads to stop logging trucks from hauling away felled timber. A few hundred kilometres away, in Eklingpura village in Udaipur, in 1988 a village leader encouraged his neighbours to form a forest protection and management committee. By abolishing grazing and controlling illicit cutting, thefts, and fires, forest species regenerated quickly from rootstock. Protected from overgrazing, grass yields soon rose so substantially that by 1990, the 198 village households were harvesting 300 metric tons annually from their protected forests.[33] Inspired by these local-protection efforts the Rajasthan Forest Department's interest in supporting community forest management groups has grown in recent years. With the passing of a state resolution encouraging village participation in forest protection, an increasing number of young officers are starting to work with communities. Given the fragile nature of Rajasthan's ecosystems, more intensive management through community supervision is essential to establish greater environmental stability.

In the south-west state of Gujarat, most of the natural forests are located along the border areas with Rajasthan, Madhya

Pradesh, and Maharashtra. Gujarat has one of the lowest per capita forest ratios in India, under 0.05 hectare per person. Since Independence, good forest vegetation with more than 40 per cent crown closure has declined dramatically and, according to land satellite imagery from 1987, covers only 29 per cent of the official forest land.[34] The remaining forest lands are in varying states of disturbance. Despite bans on felling, illegal logging continues, as well as forest fires, fuelwood headloading, and open grazing. Dams constructed in the Satpura hill tract of the eastern part of the state have also displaced thousands of tribal people, who often rely on fuelwood headloading to survive after being displaced.

As discussed in Chapter 4, a long history of conflicts between tribal communities and the forest department has prevailed, averaging 18,000 reported offences annually. Between 1985 and 1990 alone, 383 cases of assault against forestry staff were registered.[35] Many of the more recent conflicts stemmed from the ban on logging imposed in 1981, when nearly 20,000 tribals employed by Forest Labour Co-operative Societies lost their jobs as a result of the moratorium. When foresters attempted to implement the logging ban, tribals were spurred by timber entrepreneurs to fight back. State politicians also exploited the opportunity to support the tribal resistance in order to gain their votes.

In 1987, in response to the growing conflicts on forest lands in southern Gujarat, some forest officers began to seek out tribal leaders in an effort to address the alienation between the department and forest communities. While logging provided jobs for a small proportion of tribal households, most families in the area were suffering hardships from the spiralling negative impact of deforestation. Gradually, community leaders agreed to curb the activities of illegal loggers and to voluntarily participate in forest-protection activities. The Gujarat Forest Department organized a series of rallies involving a fifty-kilometre environmental aware-ness-raising trek through twenty villages. During the three-day rally, an estimated 30,000 tribal villagers participated in the events. The forest department strongly encouraged the communities to form FPCs which could take the lead in developing new access controls and micromanagement plans. By early 1990 in Surat District, about 200 committees had been formed. In areas where rootstock was healthy, village-protected lands regenerated new teak shoots and companion species. In more barren areas, enrichment planting of timber and non-timber species was done. As an outgrowth of several years of protection and visibly successful regeneration, many communities began to develop a strong sense of proprietorship and pride towards the resources, further solidifying their commitment to the task.

Spontaneous cases of community forest protection were also reported in the tribal areas of the state. In 1985, in Santrampur *Taluka* in the Panchmahals region of eastern Gujarat, tribal communities began protecting forest tracts that had been logged in the 1970s. By 1993, forty-five communities were involved in public forest protection, sometimes working as individual hamlets and occasionally through intervillage collaboration. Recently, a local NGO has become involved in documenting and supporting these groups.[36]

In 1990, the Gujarat state government passed an official resolution recognizing the rights of FPCs. The forest department has since worked increasingly well with communities in managing small patches of highly degraded forests on hills alongside the agricultural fringe areas of south-eastern Gujarat. Progress in offering management partnerships to tribal communities within or near larger, better-stocked forest tracts has been slower. Yet, in Gujarat and other Indian states natural forests that are still intact, are most in need of protection and tighter access controls. Forests rich in timber and biodiversity — whether classified as reserved, protected (demarcated or undemarcated), or protected wildlife areas — are currently the most vulnerable to accelerated degradation. Such areas should receive priority for catalyzing community-based protection.

Himalayan Zone

The Himalayan mountains rise from the Gangetic plains, first from the Shiwalik foothills, ranging from fifteen to fifty kilometres in width. Behind the Shiwaliks, the inner hills vary in height from 1000 to 3000 metres, followed by a broader middle hill belt. The greater Himalayas rise high above the forest line, with the expansive Tibetan plateau lying within the rain shadow. Most of the region's population resides in the foothills, inner ranges, and middle hills. Past traditions of exchange and use agreements among farmers, graziers, and traders have minimized resource conflicts. Given the steep and rocky terrain, survival for the hill and mountainous communities of the Himalayas was difficult. Terracing and resource nutrient transfers, based on composting leaves and organic materials from forests, allowed fertility levels to sustain acceptable levels of crop production. Forests were also essential for supplying livestock with grazing land and fodder, critical to the pastoral systems that complement agriculture. Historically, in order to maximize the ecological benefits from these forests, 'Village sites were usually chosen halfway up the spur, below oak forests and the perennial springs associated with them . . . '[37] Indeed, the sustainable management of soils, forests, water, grazing land, and other natural resources was essential.

The ecological and economic importance of forest resources has grown over time as populations have expanded, both in the Himalayas and in the broad Gangetic Plain. Both regions are vulnerable to floods and droughts affected by upstream conditions. Vegetative forest cover on the steep upland terrain, which characterizes the region, helps to stabilize run-off, reduce soil erosion, landslides, and downstream flooding. Forests also serve as a source of fuelwood energy, wild foods, grazing land, and mulch — all elements central to the subsistence household economy of millions of hill families.

Today, the once lush forests of the world's highest mountain range have been largely reduced to small patches on ridge tops and steep valley slopes. Indigenous management systems have been trammelled as the state imposed its own priorities and controls, often in conflict with village use practises. During the colonial period in the mid-nineteenth century, the state began directly controlling forest lands. Between 1815 and 1878, the government concentrated on acquiring the valuable sal *(Shorea robusta)* forests of the Shiwaliks and inner hills. The passing of the Indian Forest Act of 1878 firmly entrenched state-monopoly over much of the hill forest-lands, mandating clear procedures to enlist them under the jurisdiction of the state forest department. From 1878 to 1893, the majority of middle hill forests of Almora and Garhwal districts were designated state reserved forest. Unless a community could file and follow a protest through the court system, it was inevitable that it would have to relinquish all user-claims. Finally, in 1893, 'all wasteland not forming part of the measured area of villages or of the forests earlier reserved was declared to be DPF (Demarcated Protected Forest) under the act.'[38]

In Uttar Pradesh, hill communities agitated against the loss of their rights. This forced the government to mildly concede so that villages could retain some control of neighbouring forests under the Van Panchayat Act. However, Van Panchayats were guided by rules established by the revenue department, Panchayat directorate, and forest department. Consequently, these regulations could scarcely be interpreted as components of independent indigenous management systems. In recent years, some communities have successfully attempted to improve Van Panchayat forest management by subdividing their forests into parcels protected by single hamlets.[39]

The growth of human and livestock populations has magnified the problem of open access abuse in a management vacuum. In Tehri District, for example, the number of permanent residents increased 280 per cent between 1940 and 1981. Thousands of Nepali labourers, military personnel, and contractors flooded the area and added critical extraction pressures to an already depleting forest resource.[40] With rights and responsibilities split ambiguously between government bureaucracies and local communities, few effective controls have been actualized.

Community resistance to state-condoned logging occurred repeatedly during the twentieth century. One of the most famous struggles in 1973 came to be known as the 'Chipko' movement, a Hindi term for hugging. Cutting rights to a local forest in Garhwal had been leased to a city sporting goods company, enraging community members who had earlier been refused access to timber for making agricultural tools. With encouragement from a local NGO, villagers organized successfully to protect the forest from loggers by hugging the trees and defying the axes. Over the next decade 'tree hugging' protests spread through a number of communities in the Uttar Pradesh hills. Village-based environmental activism began to disrupt forest department felling rotations.[41] Unfortunately, the state of Uttar Pradesh has been slow to respond to grassroots protests by clearly altering its policy to empower community groups in the management of their fragile hill forests.

Healing the damaged forests of the Himalayas presents management and technical problems. Due to typically shallow and nutrient-poor soils, colder climates, and a shorter growing season, tree growth is slower than in the plains. Experimental tree-planting schemes have not been encouraging, partly because of poor plant stock and species choices, and partly because of the difficult conditions and the institutional problems inherent in undefined tenurial benefits. Natural regeneration of native species, combined with enrichment planting in some areas, may offer the most promising approach to enhancing forest productivity. This strategy will also demand close community participation, with an emphasis on establishing an equitable system of much tighter access controls than in the past.

The role of women in the biomass-based economy of hill communities is now also recognized as critical to any sustainable management system. Many young and middle-aged males migrate to the plains for extended periods during the year. Women provide a continuous presence in the communities and forests, and are the dominant collectors, users, and processors of wood and non-wood forest resources. In Himachal Pradesh, many village women's organizations *(Mahila Mandals)* have taken the lead in devising strategies to respond to social and environmental problems.

In Himachal Pradash, as well as Jammu and Kashmir, stretching from the Shiwalik hills to the Greater Himalayas, forests have played a strategic role as a buffer for agriculturalists and nomadic pastoralists. Farming and herding communities have developed mutual use agreements to reduce access conflicts and ensure long-term productivity. As discussed in Chapter 5, the nationalization of forest lands and the intervention of the state in communal land management have progressively undermined indigenous systems of use and control. Escalating population growth further exacerbates degradation as local controls erode. Despite these socio-political changes, traditional management systems, such as the village *Shamilat* forests, still operate to some extent in many areas. Yet, due to a century of competition and confrontation with the forest department, many communities are now distrustful and fear that if they invest in management, their remaining rights might also be usurped.

In Jammu, the forest department and user communities began

experimenting with village forest committees in 1988. Fifteen villages in Chinota decided to protect 350 hectares of forest land; in return, the government agreed to the formation of a village committee to oversee social forestry projects, including the selection of species for enrichment planting, record keeping, evaluation of staff performance, and nomination of candidates for forest guard positions. Community members voted to halt grazing on regenerating forest lands; in turn, the forest department, despite its policy of total closure, allowed the collection of fodder grasses. Over the next five years, 617 village forest committees were formed in the Jammu area with the active encouragement of the forest department. Unfortunately, many of these groups are inactive today. Few groups hold regular meetings and there is little involvement of women. This may reflect a lack of management training and support. Yet, it may also indicate that the forest department has inappropriately attempted to channel community participation in natural forest management through its social forestry wing, which emphasizes the protection of smaller plantation tracts. This is reflected in the small area protected by a single village forest committee, averaging only 6.2 hectares.

Ultimately, for the communities of Jammu and Kashmir to take joint forest management seriously, the programme will need to articulate clearly to the communities their rights and responsibilities for managing degraded forests and pastures. This necessitates the active involvement of the territorial as well as the social forestry wings of the forest department. It will also require examining how migratory pastoralist groups might be accommodated in future agreements. The initiation of a joint forest management resolution and programme in Jammu and Kashmir has helped in changing attitudes of the forest department field staff. Initial experience with natural regeneration over the first five years of the programme have also been encouraging. Local species of *dhak (Butea monosperma)*, bamboo, *semal (Bombax malibaricum)*, and *Dudonea viscosa* have regenerated well in many sites, often suppressing seedlings of exotic species. The impact of protection on grass productivity has been particularly dramatic, allowing some communities to change over to stall-feeding and upgrading cattle for commercial dairying.[42]

In north-eastern India, the populations of most states (Manipur, Meghalaya, Mizoram, Nagaland, Arunachal Pradesh,

Tripura, Assam) are largely comprised of tribal inhabitants with agrarian resource-use systems and cultures that are heavily forest-dependent. Conflicts over forests have been a component of tribal resistance to state intervention since the early nineteenth century. Tribal concerns over the alienation of their ancestral resources to government and outside migrants led to protests in the mid-1960s, which turned into the Naga and Mizo revolts.[43]

Historically, the forest department and tribal communities have disagreed over land use practices. Many tribal groups in the region practice rotational shifting agriculture (*jhum*), which involves the clearing of small forest plots and the burning of dried slash just before the monsoon. Crops are grown for one to two years, after which the land is allowed to lie fallow for twenty years or more before returning to the patch. While studies from Asia indicate that traditional shifting cultivation can be a highly productive and sustainable land use system given proper land recovery time, the revenue and forest departments generally condemn such practices. This has placed forest farmers in direct confrontation with the state. Unsustainable commercial logging, stimulated by the escalating demands of urban industrial centres in India and Bangladesh, has accelerated forest denudation throughout the north-east. Growing human pressures for agricultural expansion fuelled by increasing migrant flows have also reduced the pool of long rotational forests drastically, reducing the requisite fallow-period in some areas to as little as four to five years. This leads to serious losses in soil fertility and constraints on the processes of natural regeneration.[44] Nonetheless, communal systems of resource management remain stronger among the tribes of north-eastern India than in many other parts of the country. New joint forest management government resolutions from Tripura and Assam now provide a sound basis for collaborative planning and improved management. Through better access controls, illegal logging could be reduced and jhum farming regulated to ensure sufficiently long rotation periods.

The Western Ghats

The forests of the Western Ghats have been referred to as the 'backbone of the ecology and economy of South India.'[45] A source of rivers that feed agricultural plains to the east and west, the Ghats are endowed with immense biological diversity, while its forests are highly productive in timber, grasses, and leaf biomass. In the pre-British period, much of the forest land was managed by local communities. Sacred groves, often sited in the periphery of water sources, were carefully protected and monitored. Other forests were specifically designated and managed for fuelwood, grazing, and leaf manure. British colonial policy, however, clashed with local management objectives, emphasizing state control and the active extraction of teak and other timber suitable for ship-building. Over the past two hundred years, the nationalization and commercial exploitation of the forests of Western Ghats have greatly undermined community management systems and eroded the size and quality of the forest cover.

Recent surveys of community forest management systems, however, indicate that village resource-protection systems are still operating in some areas. In Shimoga and Mysore districts, an estimated 80 village groups regulate access to local forests varying in size from 1 hectare to 405 hectares. Many of these groups have formed over the past twenty years, although others have been operating for generations. Managed through cohesive committees, membership tends to be small and typically exclusive of women. The committees often serve numerous functions, managing a range of natural resources and village affairs. Knowledge about the locations and territorial ranges of local forest-management groups is limited. While these community forest-protection groups have no formal ties with the forest department, they offer great potential for improving management. NGOs active in Karnataka and Kerala have also inspired and organized villages to assume a more active role in environmental conservation. Tribal communities in the Nilgiri Biosphere Reserve and the Palani Hills in Tamilnadu have begun forming committees to address resource degradation problems. In the past, the absence of state-level policy support to empower community user group managers limited the potential for collaboration between the forest department and villages. However, the Karnataka government resolution of 1993 on joint management provides the catalytic foundation for extending the role of rural communities in forest protection.

Summary

India's population of more than 900 million continues to grow steadily at a rate of 2.1 per cent a year. Most specialists estimate that the nation's population will not stabilize until it reaches between 1.5 and 2 billion, surpassing that of China by the year 2015.[46] In view of these demographic realities, the nation's natural resource base will need extremely careful and intensive management to meet basic human and environmental needs. At present, forests, groundwater, and soil resources are being degraded and drawn down at a rapid, unsustainable rate. Highly decentralized local community protection may offer the best prospect to achieve controls that can ensure environmental conservation and productive use. Yet, given the strong emergence of bureaucratic systems of management over the past century, a major socio-political transition will be necessary to formally enlist communities in the new equation. Reversals in policy, practice, and attitude at all levels of government will be essential for stabilizing resource use.

Fortunately, many communities in India are responding to resource scarcities by developing localized protection and management systems. They are attempting to regain control over the forests and water systems central to their survival. In some areas, for centuries villagers have resisted attempts by outsiders to exploit their forest resources. In other regions, communities are organizing for the first time in response to a new awareness of their environmental crisis. Many community members have demonstrated that they can effectively protect degraded natural forests and enhance its rapid regeneration, biodiversity, biological productivity, and ecological functioning. Yet, local resource management initiatives based on small communities, often comprised of fewer than twenty to fifty households, are vulnerable to collapse, induced by more powerful individuals and groups, whether they be neighbouring villages, local politicians, business people, or state agencies. These grassroots environmental movements, interpreted as local responses to the environmental crises, must be encouraged to grow in their capacity as able forest-keepers and leaders in the rehabilitation of vast areas of wasteland throughout India. It is essential to better understand the conditions, incentives, and processes that are driving this environmental activism in order to facilitate their replication in other parts of India and the world.

Notes

1. Dan Ritchie, *A Strategy for Asian Forestry Development* (Washington, D.C.: World Bank, 1992).
2. 'Battling for Control', *Down to Earth*, 30 November 1993, p. 9.
3. Robert S. Anderson and Walter Huber, *The Hour of the Fox: Tropical Forests, The World Bank, and Indigenous People in Central India* (New Delhi: Vistaar Publications, 1988), p. 14.
4. 'Global Deforestation,' *New York Times*, 6 July 1990; Anil Agarwal and Sunita Narain also estimate forest loss to be about 1.5 million hectares annually, in their book *Towards Green Village* (New Delhi: Centre for Science and Environment, 1989), p. 1.
5. SPWD (Society for the Promotion of Wastelands Development), 'Joint Forest Management: Concept and Opportunities' (Proceedings of the National Workshop at Surajkund, New Delhi, August 1992), p. 3.
6. Mark Poffenberger and Betsy McGean, 'Policy Dialogue on Natural Forest Regeneration and Community Management', Research Network Report no. 5 (Honolulu: East-West Center, April 1994), p. 9.
7. J.B. Lal, 'Economic Value of India's Forest Stock' in Anil Agarwal (ed.), *The Price of Forests* (New Delhi: Centre for Science and Environment, 1992), p. 44.
8. International Development Association, 'Report and Recommendations of the President of the Executive Directors on a Proposed Credit to the Government of India for Forestry Technical Assistance' (Washington, D.C.: World Bank, October 1975).
9. See Dan Ritchie, *A Strategy for Asian Forestry Development.*
10. B.B. Vohra, 'The Greening of India' (New Delhi: Intach, 1985), pp. 14–15.
11. Ibid., p. 2.
12. Ibid., p. 13.
13. Ibid.
14. See R.C. Verma, *Indian Tribes Through the Ages* (New Delhi: Ministry of Information and Broadcasting, GOI, 1990), pp. 225–28.
15. Ibid., pp. 92–101.
16. See B.B. Vohra, 'The Greening of India', pp. 13–14.
17. Nadeem Hasnain, *Tribal India Today* (New Delhi: Harnam Publications, 1991), p. 155.
18. K.C. Malhotra, 'Role of Non-timber Forest Produce in a Village Economy: A Household Survey in Jamboni Range, Midnapore District, West Bengal', Working paper (Calcutta: IBRAD, 1991).

19. Mark Poffenberger and Samar Singh, 'Forest Management Partnerships: Regenerating India's Forest', *Unasylva* 170(43): 46, 1992.
20. Personal communication from Madhu Sarin, February 1994.
21. Mike Arnold, Ruth Alsop, and Axel Bergman, *Evaluation of the Sida-Supported Bihar Social Forestry Project for Chotanagpur and Santhal Parganas, India* (New Delhi: Swedish International Development Authority, March 1990), p. 11.
22. Personal communication from Madhu Sarin, February 1994.
23. Shivnath Mehrotra and Chandra Kishore, 'A Study of Voluntary Forest Protection in Chotanagpur, Bihar' (Bhopal: Indian Institute of Forest Management, 1990), pp. 1–2.
24. Personal communication from Madhu Sarin, February 1994.
25. See Shivnath Mehrotra and Chandra Kishore, pp. 31–2.
26. Amit Mitra and Kanti Kumar, 'Death by Starvation,' *Down to Earth*, 15 June 1993, p. 32.
27. Shashi Kant, 'Gandhian Approach to the Management of Forest as Common Property Resource: A Case Study of Binjgiri Hill (Orissa), India' (paper presented at the First Annual Meeting of the International Association for the Study of Common Property at Duke University, Durham, N.C., 27–30 September 1990), p. 5.
28. Ibid., p. 6.
29. SIDA, 'Helping Forest Dwellers of Orissa to Adopt Viable Alternatives to Shifting Cultivation' (Bhubaneshwar: Center for Development Research and Training, 25 December 1990).
30. Orissa Forest Department, unpublished Area Statement, 1993.
31. See note 29.
32. See SPWD 'Joint Forest Management: Concept and Opportunities', (Proceedings of the National Workshop, 1992), pp. 22–3.
33. Ibid., pp. 18–19.
34. R.S. Pathan, N.J. Arul, and M. Poffenberger, 'Forest Protection Committees in Gujarat: Joint Management Initiative', Sustainable Forest Management Working Paper no. 7 (New Delhi: Ford Foundation, 1990), p. 1.
35. Ibid., p. 3.
36. Personal communication from Madhu Sarin, February 1994.
37. Ramachandra Guha, *The Unquiet Woods: Ecological Change and Peasant Resistance in the Himalayas* (Delhi: Oxford University Press, 1989), p. 28.
38. Ibid., p. 44.
39. Personal communication from Mr. Kanailal, CHIRAG (an NGO).
40. Neeru Nanda, 'Local Management of Community Property Resources in the Himalaya', in Anil Agarwal (ed.), *The Price of Forests* (New Delhi: Centre for Science and Environment, 1992), pp. 254–5.
41. Madhav Gadgil and Ramachandra Guha, *The Fissured Land: An Ecological History of India* (New Delhi: Oxford University Press, 1992), pp. 222–4.
42. Jaya Chatterji and Viren Lobo, 'Study of Village Forest Committees of

Jammu Region of the State of Jammu and Kashmir' (New Delhi: Society for the Promotion of Wastelands Development, 1991).

43.   See R.C. Verma, *Indian Tribes*, pp. 225–8.

44.   P.S. Ramakrishnan, '*Jhum:* Is There a Way Out?' in Anil Agarwal (ed.), *The Price of Forests* (New Delhi: Centre for Science and Environment, 1992), pp. 304–5.

45.   Madhav Gadgil, 'Western Ghats: Managing the Forest Cover' (paper prepared for the Indo-Pakistan Conference on Environment, Lahore, 13–15 December 1989), p. 1.

46.   Barbara Crossette, 'Population: A Runaway in India', *International Herald Tribune*, 17 September 1992.

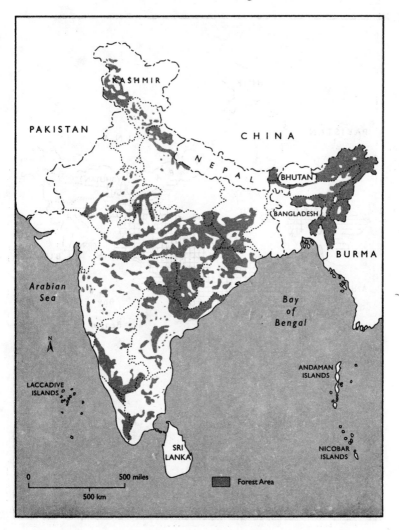

FIGURE 1.1   India's forests

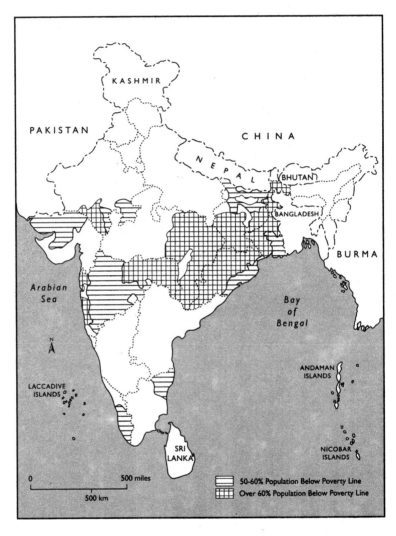

FIGURE 1.2 Poverty areas in India

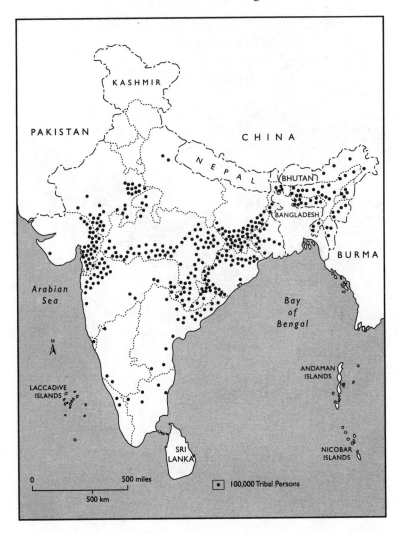

FIGURE 1.3 India's tribal concentration areas

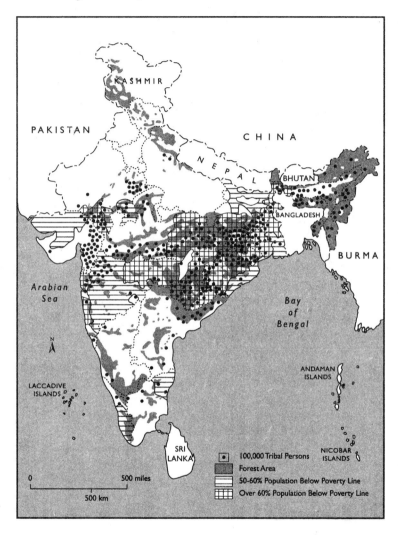

FIGURE 1.4 Forest, poverty, and tribal areas

Table 1.1 Orientation of social forestry and joint forest management programmes

Social forestry	Joint forest management
1. Objectives: – satisfy local needs through fuelwood plantations to divert pressure from natural forest – supply timber, poles, and pulpwood to industries – establish fast-growing tree plantations on 0.4 million hectare of degraded land annually	1. Objectives: – meet local needs equitably for diverse range of forest products through natural forest regeneration under community protection – extend authority to communities to control forest access and allow local management – regenerate 30–50 million hectares of degraded underproductive forest land – manage for biodiversity, ecological sustainability, and environmental benefits
2. Who: – 'communities' through village panchayat government structure (ave. 500–2000 households) – private farmers (especially larger farmers with credit access)	2. Who: – clearly defined and organized formal and informal community user groups (ave. 10–100 households) supported by the forest department – focus on most forest-dependent women, tribals, landless

(cont.)

Table 1.1 (*cont.*)

Social forestry	Joint forest management
3. Where:	**3. Where:**
– private lands	– state forest lands (protected and reserve)
– common property (revenue lands, village grazing/panchayat land)	
4. How:	**4. How:**
– forest department organized	– community organized
– making budget, setting targets	– emergence of community concern and ability to act
– establishing nurseries and plantations	– diagnosing social and ecological opportunities
– providing employment	– defining rights and responsibilities (products, benefit-sharing, protection)
	– microplanning process (access controls, silvicultural operations to enhance natural regeneration)
	– legitimizing authority of community management group

5. When:
- based on process of community activism and interest
- expansion based on spontaneous or encouraged spread to other villages

6. Average Cost: Rs 0–500 per hectare

5. When:
- based on donor aid and budget process
- renewal based on target achievements

6. Average Cost: Rs 5000–10,000 per hectare

Chapter 2

Communities and the State: Re-establishing the Balance in Indian Forest Policy

Mark Poffenberger ■ Chhatrapati Singh

After a century of policies and legislation that tightened state control over India's forests, government directives more supportive of community resource rights and responsibilities have begun to surface in recent years. The 1988 National Forest Policy proposed the creation of a 'people's movement' to protect forest resources. The June 1990 circular supporting joint forest management (JFM) initiatives, as well as earlier and subsequent state-level government orders, provided specific guidelines for the recognition of community forest protection activities. While often criticized for their legal modesty, these policy acts provide a growing legitimacy for community management of public lands. Perhaps more importantly, these resolutions are part of a larger process that is creating a conducive environment, allowing communities to take a larger role in the management of state lands. Yet, many questions have been raised regarding the implications of these new pronouncements. Do they mark a historic swing towards greater decentralization and de-bureaucratization of resource management systems, or do they represent an attempt by the government to control local communities' acquisitions of public forests? Are the institutional arrangements, tenurial provisions, production systems, and sharing agreements outlined in these government orders compatible with the emerging expectations and forest management systems being developed by India's rural villages? How might conflicts that exist between newly developing governmental concepts of joint management and local community forest protection strategies be resolved?

Answers to these questions are required to adapt and design JFM policies and programmes to better mesh with the diverse approaches emerging in rural India to stabilize and sustain forest use. Lessons can be drawn from experiences with social forestry programmes over the past two decades as well as from recent grassroots forest protection movements. Ideally, both national and state government policy directives need to support local attempts to stabilize and regenerate degraded forests and facilitate the socio-political changes that empower these emerging community institutions. This chapter explores the evolution of forest policies and some practical implications of the new government orders.

A Brief History of Indian Forest Policy

Two thousand years ago, as much as 85 per cent of the Indian subcontinent was covered with forests.[1] The forest ecosystems varied widely, as did their flora and fauna and the communities that inhabited them. While there is little written record describing early human-forest interactions, Vedic literature indicates that forests were held in high esteem, and the ethnobotanical knowledge of the people of those times was extensive.[2] One couplet from the *Rigveda* (x, 146–6) succinctly expresses the importance and reverence in which the forest was held: 'I have praised the Queen of the Forest, Mother of Wildlife, Redolent of balm, sweet scented, possessing much food . . . '[3]

Recent discoveries by environmental historians indicate that much of the earth's tropical and temperate forests were extensively manipulated by human populations. Fire, cultivation, and the planting of useful species all shaped the forest to respond to human needs. Rights and rules of access evolved over time to reduce conflicts among users. Even the most powerful rulers recognized and respected the importance of forests for communities and for the environment. Kautilya, the famous authority on statecraft in the Mauryan period, wrote a treatise on forest regulations. Shivaji, the dynamic Maratha leader, in his edict of 1670 instructed his officers that mango and jackfruit trees 'must never be touched . . . Our people have nurtured them like their own children over long periods. If they are cut, their sorrow would know no bounds . . . Rather it would bring ill repute to the ruler who hurts the citizenry. Furthermore there is grave danger in the loss of tree cover.'[4]

Early rulers who were interested in securing timber reserves and controlling territory often claimed large areas of forest. Forests were also places for occasional hunting and recruiting tribal dwellers into the army. Yet, if the rulers made excessive demands in tribal forest areas, they often met with guerilla resistance, a pattern of attack and hide, and rarely worth the cost of pursuing (see Chapter 5). Instead, the rulers from the Ashoka period, through the Gupta and Mughal periods, often left forest dwellers in peace, concentrating their political aspirations on the fertile agricultural plains and their more populous villages, towns, and cities.

It was not until the arrival of the British colonial interests in the late eighteenth century that the perceptions of the state towards the forests began to change. European tenurial concepts held forest lands to be crown land, stemming from Regalian Doctrine.[5] From 1770 to 1860, forests were increasingly viewed as an asset of the state with great commercial potential.[6] Yet, during that period no public agency was in a position to monitor or regulate its use. Gadgil cites Munro who, in 1838, complained that the teak forests were thrown open 'to all who wish to cut . . . They cut indiscriminately all that comes in their way; any range of forests . . . depriving future generations of the benefit they now enjoy.'[7] This was the first period of accelerated deforestation in India.

With the establishment of the Indian Forest Service under the Government Forest Act of 1865, an agency was established to initiate more systematic and efficient planning. Initially, nationalized lands were primarily controlled by the revenue department which was interested in converting forests to income-generating farmlands. Settlement officers from the revenue department fanned out across the country, conducting surveys and compiling lists of taxable agricultural lands and clarifying the rights of graziers and forest users. Yet, colonial forest administrators were increasingly dissatisfied with the Act of 1865 due to its restrictions in dealing with communities and their forest use practices. The Act clearly stated that the law should not 'abridge or affect any existing rights of individuals or communities' to forest lands.[8]

As Ramachandra Guha describes in Chapter 3, a heated debate existed during this period between Dietrich Brandis, the first Inspector General of Forests, and B.H. Baden-Powell over the role of government in forest management and the rights of forest communities. Brandis, representing the 'populist' camp, contended that forests should be left under the management of communities, both for reasons of social justice and for the practical purpose of avoiding social insurrections that could result from the usurpation of forest lands by the state. Baden-Powell, arguing for the 'annexationists,' viewed the forests as open access resources, with communities having neither the rights nor the capacity to manage them.[9] Ultimately, Baden-Powell's position prevailed, and in 1878 a new forest act was passed limiting private property only to continuously cultivated land. Human resource-use practices such as grazing, product collection, and temporary or rotational swidden farming were rejected as a basis for ownership, even if taxes had been paid. The 1878 Forest Act vested the Indian Forest Service with greater power to oversee forest-use, while emphasizing commercial forest management.

During the 1880s and 1890s, the forest department began a concerted effort to demarcate forest land, reserving the tracts with the greatest commercial potential. By the 1890s large reserves had been defined and huge quantities of logs were being felled and used for sleepers (crossties) needed for the construction of railway lines. Communities were given three months to contest the reservation, once the forest settlement officer had declared the state's intention to nationalize the lands. If the villagers failed to file a claim within this period, their rights were generally permanently revoked. Illiterate villagers were often unaware that a survey and demarcation process was in progress. Further, most rural villagers, especially tribal communities, had little experience with legal procedures for filing cases in courts. While their 'rights' had some protection and channels of legal recourse were available, it is hard to imagine remote tribal Indian communities of the late nineteenth century being able to avail themselves of those channels. As one legal scholar notes, 'Given the level of illiteracy and unsophistication of many tribal communities, the rules of notice, appeal and settlement have provided only the slimmest protection against the arbitrary extinction, diminution or reallocation of rights.'[10] Further, the 'rights' to forest access and products entitled to communities under the Forest Act of 1878 were eroded as time passed. For example, a departmental resolution issued in 1890 stated, 'The privileges conceded are intended to be exercised as a matter of favour and not of right, and are liable at any time, at the pleasure of the Government, to modification, curtailment or discontinuances.'[11]

In short, throughout the second half of the nineteenth century, the forests of rural communities were continuously being reserved and nationalized, while the rights of villagers were eroded through a series of legal actions. Village forest rights were often changed to privileges at the discretion of local bureaucrats. A century later, a Government of India survey found that few tribals were aware of their legal rights.[12] Knowledge of forest rights and legal procedures present in rural communities tended to be in the hands of economic elites and landowning families who used that information to their own advantage.

Both World Wars I and II drove deforestation, as huge demands for timber forced the modification or abandonment of working plans to achieve higher production. After 1947, in newly independent India, the intensity of commercial exploitation increased, partly due to the dissolution of the small rulers and landlords (zamindars) whose lands were taken over by the state governments. As these private forests were nationalized and brought under the management of state forest departments, pervasive efforts were made to exploit them before they were subjected to governmental controls. As a result of these changes in control, over twenty million hectares of forests were either logged or converted to agriculture throughout India, since Independence.[13] Strengthened ties between politicians, business people, and some foresters which were based on mutually remunerative industrial extraction, also drove deforestation, while further eroding the rights of forest communities.[14] Throughout the 1950s, 1960s, and 1970s millions of hectares of forests were leased out to business people at heavily subsidized rates. In return, they contributed generously to the coffers of politicians, who in turn put pressure on professional forest administrators to continue to allow unsustainable use practices. Such practices are common in many countries, including the United States, where recent exposés indicate that 'political pressures to maximize timber production are so inbred with the industry that it invites and sometimes colludes in massive fraud and theft.'[15]

Even after Independence, community forest users were perceived by the government as the driving force behind deforestation. Ancestral rights and usufructs granted under earlier colonial agreements and settlement arrangements were viewed, in some quarters, as overly generous 'concessions and privileges.' According to the National Forest Policy of 1952, 'The accident of a village being situated close to a forest does not prejudice the right of the country as a whole to receive the benefits of a national asset.'[16] As Lindsay notes, 'This disapproval of the supposed laxity shown to local uses of forests reverberates through numerous government reports in the 1970s and 80s, each of which insists that a tightening of the "concessions and privileges" granted to rural populations is essential to protecting the "national interest".'[17] In an attempt to increase government control over forest resources, rather than allow communities to manage and harvest products under *nistar* and other traditional rights, many forest departments established 'depots' where villagers could receive fuelwood, poles, grasses, and other produce at subsidized rates. Though still in existence, this system has been widely criticized for the corruption and inefficiencies that accompany it. Perhaps more problematic is that depots exclude rural resource-users from managing the mode of production, undermining their existing rights, responsibilities, and incentives to increase productivity of the resource base.

By 1980, nearly 23 per cent of India's total land area was under state management while the rights and management roles of an estimated 300 million rural resource users had become increasingly vague. Planners were concerned with the growing poverty, increasing population pressures, and shrinking forests. While nearly a quarter of India was designated forest, perhaps only half of that had reasonably good standing forests. In 1980, the Forest Conservation Act was passed to emphasize the social and ecological importance of these critical resources. The Act placed controls on logging, while also recognizing the needs of communities. However, it did little to reverse the historic swing towards bureaucratic control. In fact, environmental concerns often imposed further limitations on community rights, especially in wildlife zones.

As government agencies exerted greater controls over forests, millions of rural inhabitants throughout India, who had used these lands to meet basic needs of food, fuel, building materials, fibres, and medicines, increasingly lost access rights. As the rights of rural communities were eroded, conflicts grew between the state agencies and Indian villagers. Disagreements over management priorities led to unsustainable patterns of forest exploitation and to a gradual degradation of India's vast forests. By 1990, less than 10 per cent of the country possessed good forest cover.

Emerging Joint Forest Management Policies

Over the past few years, planners and forest administrators began developing new strategies to reduce the conflict between state agencies and rural groups. Attempts were made to facilitate the emergence of collaborative forest management systems that responded to national needs and local resource requirements. The Indian National Forest Policy of 1988 and the 1 June 1990 Joint Forest Management Resolution, combined with some sixteen state-level resolutions and government orders, began reshaping the policy environment, acknowledging the need to give greater rights and authority to community groups.

The national guideline to support joint forest management (JFM) begins by stating that:

The National Forest Policy, 1988, envisages people's involvement in the development and protection of forests. The requirements of fuelwood, fodder and small timber such as house-building material, of the tribals and other villagers living in and near the forests, are to be treated as first charge on forest produce. The Policy document envisages it as one of the essentials of forest management that the forest communities should be motivated to identify themselves with the development and protection of forests from which they derive benefits.

New state and national guidelines provide a formal basis to clarify the balance between the rights of the government and those of the community. If carried out by forest departments with sincerity and generosity, JFM represents a process of decentralized empowerment benefiting some of India's most disadvantaged groups, while possessing the potential to reverse forest degradation. At the same time, some fear that JFM may provide an opportunity for forest departments to strengthen their control, co-opting NGO and community-based grassroots attempts to protect forest resources. They point to the legal modesty of the wording of the resolutions and their reluctance to grant greater rights to forest protection groups, especially where forests are healthy and valuable. As one review of JFM policies noted: 'Forests remain under the proprietorship of the state, and access to and use of the forest by local communities is, when all is said and done, by the good graces of the state', rather than flowing from a recognition of latent community rights.[18] Others contend that after more than a century of growing national control, it is unrealistic to expect the government to relinquish all authority to communities and for state forest departments to disband. In fact, it may be critical that the government generally, and the forest departments specifically, continue to monitor the use of public resources. Since economically weak rural communities are pressurized by more powerful commercial interests, the forests they protect could be subject to manipulations that are inconsistent with local community or broader societal needs. This perspective suggests that decentralization is inevitable, being demanded from below by an increasingly politically powerful rural population, driven by growing resource scarcities. While the content of these initial resolutions may be disappointing, they could initiate a process of change in policy, that will shift the balance of resource control in a more equitable manner over the next one or two decades.

Of more immediate importance is the question: Do these policies support the types of spontaneous community forest management initiatives, based either on traditions or emerging need, which are increasingly present in India, or do they undermine them and erode existing rights? Due to limited experience with JFM, early resolutions formulated to support these new programmes may not always be consistent with the field realities of the programme or community needs, and in some areas may actually hinder rather than help village management initiatives. Essentially government policies and programmes must support village efforts if this promising collaboration in resource management is to succeed.

The passage of the 1989 West Bengal resolution may be seen as a precedent in the shifting orientation of Indian forest management policy. Its significance lies in the context in which it was passed. West Bengal forest officers had begun experimenting and developing unwritten policies since the early 1970s by encouraging individual communities to take over protection responsibilities for public forest lands. Given the populist socio-political environment in West Bengal during this period, many villages had also been acting on their own to control forest access and assert their exclusive management rights in relation to other neighbouring communities. The West Bengal Forest Department resolution formulated an official programme and provided government legitimacy to this grassroots resource management movement.

Due to its influence on subsequent national and state resolutions supporting JFM in India, certain components of the West Bengal document are noteworthy. Broadly, the document provides guidelines for management group members, the structure and functioning of the executive body, protection responsibilities, and usufruct rights. The resolution also attempts to clarify the relationship of small informal village groups to the local government by placing them under the supervision of the land use committee (*Bon-O-Bhumi Sanskar Sthayee Samiti*) of the elected body of sub-district level representatives (*Zilla Parishad*). The document requires that Forest Protection Committees (FPCs) hold annual meetings, elect representatives, and maintain record books. The policy states that the primary function of FPCs is to ensure the protection of the forest consistent with existing forest policies. In return each FPC is entitled to a 25 per cent share of timber produced in its area and to all non-timber forest products. The resolution includes a clause mandating a five-year vesting period during which protection must be effectively carried out prior to the granting of rights in the timber harvest, though non-timber forest products are available from the outset of the agreement. The resolution concludes with a statement empowering the forest department and the Zilla Parishad to dissolve FPCs that fail to perform their protection responsibilities.

While the West Bengal resolution lacks clarity in some areas and also raises some problematic institutional and economic issues, it is, perhaps, most significant for its authority to provide some legitimacy to more than 2000 community forest management groups currently protecting 300,000 hectares of state forest lands. To that end, it provided *de jure* status to a *de facto* management system.

Encouraged by the extensive and generally successful experiences with JFM in West Bengal, a group of individuals from both government and non-governmental organizations (NGOs) drafted a national resolution to encourage state forest departments in other parts of India to experiment with collaborative-management systems. The resolution, which was approved in June 1990, encourages NGOs, state forest departments, and community groups to collaborate in managing state forest lands. The resolution gives considerable latitude in the types of community groups that might be involved, specifying panchayats, co-operatives, and informal village organizations. It notes that no restrictions should be placed on membership. The guidelines suggest that communities should be given some rights to timber and non-timber products from jointly-managed community lands, referring to the 25 per cent share of timber allocated by the West Bengal Forest Department to participating community groups. It also recommends that community management groups be involved in the formulation of working plans for the forest area under their protection. While the national resolution suggests that the community must protect the forest from illegal extractive activities and ensure that no grazing or agriculture is carried out on forest lands, it also encourages fruit tree, shrub, and grass enrichment planting.

In addition to the national and West Bengal resolutions, all Indian states except Uttar Pradesh, Assam, Manipur, and Nagaland have approved JFM guidelines. Selected state resolutions are summarized in Table 2.1. These resolutions possess similar provisions for recognizing communities as management partners for specified tracts of government forest lands. Although most state resolutions allocate all non-timber forest products for community use, timber-sharing arrangements vary from fuelwood in Bihar, timber for subsistence-use in Orissa, and up to 60 per cent of net commercial timber revenues to participating communities in Rajasthan. In terms of the structures recommended for community management organizations, the resolutions also vary. In Orissa, the panchayat is suggested as the management unit, while the Bihar resolution suggests village development committees. In both Haryana and Rajasthan, the government orders indicate that community-based Forest Protection Committees become formal bodies through registration under the Indian Societies Act. While all state resolutions suggest that community management groups take the responsibility for forest protection, some states require that management groups take a specified form in terms of committee membership, leadership, meetings, and record-keeping. Both national and some state resolutions clearly note that no ownership or lease rights are implied by JFM programmes implemented under these guidelines. However, the national resolution does state that participating communities should be involved for a ten-year period with the option to renew their management agreements with the forest department. Similar provisions are outlined in the West Bengal resolution.

To summarize, the national and state notifications share a number of common features. These include the following:

- All resolutions provide user groups with usufruct rights only. They clearly note that land is not to be allocated or leased.
- The resolutions generally recommend village-level committees as functional management groups.
- Village-level groups are to operate under the supervision of forest department officials.
- In most resolutions the tenurial period for usufruct rights is not mentioned.
- Many resolutions limit community management and benefits to degraded forest land.
- Some resolutions propose the formulation of joint management plans to co-ordinate agency and community activities. However, operational guidelines for co-operative planning are unclear, and where procedures are outlined, they tend to reflect forest department interests and concerns.

Forest Management Policy and Operational Issues

Newly drafted national and state-level resolutions supporting the establishment of JFM systems in India provide an attractive opportunity for experimenting with community-forest department collaboration. These guidelines will, however, need further improvement as awareness grows (i.e. ways to encourage JFM activities by supporting the development of local organizations, facilitating women's participation, enhancing tenurial security, and providing appropriate economic incentives).

The Role of Local Government

Researchers studying the process of community forest management group formation have found that a great deal of grassroots organizing for environmental management is taking place in various parts of India. Where state forest departments are supportive, village forest management groups are often able to sustain protection effectively, even when under pressure from other communities and the private sector. With the approval of state JFM resolutions, these groups can be given some formal legitimacy which would further strengthen their authority. Important policy questions include the form local organizations should take, how they might be recognized, and their legal status.

Many of the forest protection groups observed in both western and eastern India are comprised of ten to fifty households in a small hamlet (e.g., sahi, phalia). Generally, because of their small size, these groups are more homogenous than the larger panchayat and may be comprised of one or two castes, tribes, or clans. The resulting social and economic uniformity of the group has facilitated group decision-making and action. If these forest management groups are formed or exist at the sub-village panchayat level, what is their relationship to the local government? There is some concern that these emerging forest management organizations may not fit well within the system of local governance, including newly evolving (Panchayati Raj) systems, and that linking them to panchayats may undermine their independence and effectiveness.

Most resolutions suggest that local panchayat institutions monitor the activities of village groups, as is the case in West Bengal. The Orissa resolution notes that the forest protection committees (*Vana Samrakshana Samiti*) should be placed under the supervision of the 'Sarpanch of the concerned Gram Panchayat and the Forester (who) shall be the Chairman and Convenor of the committee respectively.'[19] A subsequent resolution states further, 'There should be one Samiti for a group of hamlets or villages if these are all located adjacent to the forest area to be regenerated.' This implies extending membership in forest management groups to all communities in the area and placing them under the leadership of the local panchayat headperson. However, in Orissa a panchayat may cover up to twenty villages.

In such cases, to officially comply with the resolution, the community that originally formed a forest-protection group would have to join other villages in creating a new management system. In the process, the original community's authority over forest management decision-making would be eroded. Where existing community management groups are functioning, they would be obligated to share forest-produce with other communities incorporated into the management committee. Most community forest-protection groups desire a clear definition and absolute rights to the forests they protect to ensure that the benefits from their protection activities flow directly to them. Consequently, merging small hamlet-based groups could undermine the rights of responsibility and use.

There is also some concern that if JFM groups were absorbed by a village panchayat, vested interests that continue to be influential in many communities might exert control over decision making. While more democratically elected village panchayats are developing in Karnataka and West Bengal where the new Panchayati Raj system has been adopted, in many states traditional elites still effectively manipulate village panchayat decision making. Since many small community-based forest protection groups are comprised of tribals, scheduled castes, and less powerful groups and communities within the larger panchayat, they may lose their authority to the elites if the management groups become a direct adjunct of the panchayat.

Experience from Indian social forestry programmes over the past twenty years indicates that in many cases panchayats have difficulties in effectively managing community woodlots. In some cases, due to their inherent political nature and often diverse constituencies, panchayat-based management groups had difficulties reaching a consensus regarding the management of community forest resources. Case studies from India and Nepal have demonstrated that left on their own, villagers often organize around small community groups in physical proximity to the forest, allowing them to effectively mobilize to establish management systems including protection activities, harvesting and sharing systems, and dispute arbitration. There is a need to clarify the relationship of local forest management groups to local government institutions. At the same time, simply subsuming them into the local government body could threaten their effectiveness.

While any rigid attempt to force small community forest management groups to merge to better conform to local governance structures might undermine their effectiveness, experience from Orissa and West Bengal indicates that some forest protection groups are forming apex organizations. For example, in Nayagarh District in Orissa, 365 villages are protecting an estimated 86,000 acres of state forest. These communities have formed a co-ordinating body (*Mahasangh*) to represent their interests before the state government and the forest department and to resolve their conflicts.[20] In some cases, elected officials (members of the legislative assembly) assist in organizing apex groups and actively lobby in support of the interests of forest-protection groups at the local-government level. Field documentation of such situations indicates that the component communities generally agree to keep their independent identity within the larger group by maintaining clear boundaries of their area and retaining exclusive control over harvests in their territory. The function of the larger umbrella organization is often to facilitate the expansion of joint-protection activities through organizing new communities, resolving inter-village disputes, and assisting member committees in dealing with the local government and the forest department. In some cases, such apex groups have protested the structural prescriptions and product-sharing guidelines set forth in state resolutions.

It may be more useful to explore the role local governments could play in working collaboratively with the forest department in assisting with planning and monitoring forest management activities of local groups within their jurisdiction. Some past state resolutions, including those from West Bengal, are very specific and direct regarding the role of the forest department and local government in determining committee members. The original 1989 resolution stated, 'The beneficiaries shall be identified from amongst the economically backward people living in the vicinity of forests concerned', and that the local panchayat land management committee 'shall select beneficiaries for construction of the forest protection committee.' This statement implies that local government representatives from outside the community would determine who could and could not participate. In 1990, the West Bengal resolution was revised to allow every family in the village to be a member of the management group. However, the new resolution maintained the clause that the panchayat land management committee and the West Bengal Forest Department should determine eligible families to benefit from the programme.

This raises a fundamental question regarding whether government agencies should intervene in determining the membership of community resource management organizations. To do so would erode community independence — making FPCs simple creations of bureaucratic agencies — and its local legitimacy and ability to carry out management functions. If membership is selective and determined by outside agencies and local government bodies, some families already participating in JFM activities might be denied membership and would be excluded from the programme. Further, it seems fundamental that community-based organizations have the final authority over their own membership. Throughout eastern India, most village management groups were organized by the community alone, whose members determined their own membership. In a survey of FPCs in West Bengal, those comprised of most or all households in the community had more effective forest management organizations, in terms of holding regular patrols and meetings, than those with partial membership.

In areas where local groups function effectively, the forest department and the local government may only need to formally empower them and provide technical assistance. At an operational level in the field, this practice is generally occurring in West Bengal. To be supportive of grassroots initiatives, JFM policies should reflect flexibility in encouraging the diversity of institutional strategies that rural communities use to organize, rather than providing prescriptive models or 'blueprints' that must be followed. Forest department officials and policymakers need better information regarding how and why communities form organizations to manage forest resources, and how they might best relate to local government bodies. This knowledge would evolve in clearer policies to support these emerging and promising community organizations.

Registration and Dissolution

Recent resolutions can drive forest departments to hurriedly establish community resource management groups in a top-down manner, undermining their own self-identity. In 1988 in Orissa, a resolution was issued to form FPCs. The Minister of Forests decreed that 5000 groups were to be established during the remaining six months of the year. While forest department officers successfully registered hundreds of existing community forest management organizations, they also rapidly formed several thousand new groups, most of which failed to develop management functions.[21]

The existence of community organizations that are solely formed by and dependent on forest departments, undermines the growth of independent local self-governance and the development of democratic institutions and processes. Community forest management groups have little say over policy and management decisions. Most state resolutions allow the forest department to dissolve forest management organizations if they fail to perform as expected by the department. While forest departments may insist on an 'escape' clause in the resolution to negate the management agreement if their community partners fail to uphold their responsibilities under the JFM programme, it is also important that the separate identity of village resource management organizations be respected.

In Rajasthan and Haryana, where the resolutions require forest management groups to become registered societies, protection committees have a separate and legal identity and, consequently, greater independence. Yet, the registration process is complicated, as the Societies Registration Act was not originally designed to meet the needs of village organizations. In Haryana and West Bengal some committees have requested assistance from MLAs and other politically elected leaders to strengthen their bargaining power. In the Pinjore area north of Chandigarh, fourteen community management groups requested the Haryana Forest Department to modify the terms of the grass lease-pricing and payment system. These experiences suggest that community forest management groups will want to maintain a separate identity and utilize local governance bodies, elected leaders and group apex organizations to express their needs and negotiate more effectively with the forest department. The needs for autonomy and democratic process at the community level are currently not reflected in state resolutions but should be given careful consideration when these documents are revised.

Women in Forest Management

Most resolutions lack clarity regarding the role of women in JFM systems. While the Gujarat resolution specially requires at least two women members on community management committees, and revised guidelines for the JFM programme in Haryana require both male and female household heads be members of the community forest management society, most resolutions do not refer to women's participation. Since women are frequently the primary users of forest resources, they play an important role in management decision-making and are formally recognized as voting members of local management groups. Further, in some rural areas where men migrate from the village for extended periods or are too busy with agricultural work to allocate time to management activities, it may be appropriate for the community to establish management groups comprised solely of women. In fact, this has already taken place in a number of states. Resolutions need to be written so that they have the flexibility to support emerging community initiatives and not assume a single model of community management.

Women's involvement in JFM, however, cannot be simply legislated. Communities should perceive its importance and male members will need to support it. In many cases, the work of the forest protection committee is entirely dictated by men. A 33-year-old woman in Paraswara village in Madhya Pradesh notes that for her the 'FPC meant a meeting of men'.[22] In Orissa, in one village studied by researchers, the FPC met in a meeting hall where women were not allowed.[23] Even where women are formally represented on FPCs, they are often too shy to speak before the gathered men. Since virtually all field-level forest department staff are men, communication taboos between men and women also limit discussions. Consequently, JFM programmes suffer from gender inequities present in rural Indian society. Until rural women are granted status and greater opportunities, the role of females in the management of forests, even within JFM programmes, will be constrained.

Tenurial and Use Rights

In many forest areas in India, communities and panchayats historically possessed a range of customary legal rights over forest resources (e.g., nistar, *dafavati*) which have been recognized under earlier forest acts. In the past decades, national and state governments have reduced these rights to revocable 'privileges.' If communities are to have the authority and confidence to protect and manage forest lands, their rights need to be clear and well established. This requires a reversal of the government's past policy in extending its domain and authority, by granting greater powers and rights to the village managers.

Under the emerging joint management programmes, forest departments can establish new agreements with communities and extend new rights to local groups. This process can, however, create problems if earlier rights-holders are excluded from the new agreements or pre-existing rights are not recognized. In India, there are still large areas where forest rights have not yet been settled. Customary rights should be settled prior to formalizing agreements with forest management groups. Rapid-appraisal field documentation and micro-planning exercises could facilitate this process. Ultimately, communities are often best able to work out usufruct agreements. Bureaucratic agencies often have little detailed information on local resource use patterns and perceived access rights. Forest departments may be most effective in creating opportunities for neighbouring villages to reach resource-use agreements among themselves and ratify their decisions. Many of the FPCs operating in south-west Bengal negotiated with neighbouring communities to clarify rights and territorial responsibilities when they began to initiate protection activities. In many cases, these communities have the strongest incentive to avoid conflicts with their neighbours over forest-rights and access.

In eastern India, many villages have demonstrated that they can conduct much of the negotiation on their own or with the assistance of the local government. The forest department, however, holds ultimate responsibility for ensuring that its agreement with management groups does not create conflicts over pre-existing, real or perceived, usufruct. Further, once an inter-village consensus has been reached regarding forest management rights, after a period of time such agreements need formal, legal approval in some form.

Some government agencies are also empowered with certain rights and interests. Panchayats and parastatal marketing organizations (LAMPS and Trifed, as well as local co-operatives) often have harvesting and marketing rights to timber and non-timber products, which must be altered or recognized when the forest department is developing usufruct agreements with community management groups.

Except for clauses in the national and West Bengal resolutions, most state programme guidelines do not provide long-term rights to communities which protect and hope to benefit from forest lands under joint management. Clear tenurial security enhances the authority of community management groups to carry out protection activities, especially when under pressure from neighbouring villages and private interest groups. Participating communities, which invest labour in protection activities and defer exploitation of forest resources to benefit from future production increases, may need greater assurance of the government's commitment to the programme. Since state forest departments are in essence entering into management partnerships with village groups, the time frame for such agreements, as well as the basis for extensions, must be clearly defined. Without a clear time mandate, community management groups may fear that their labour investments will not yield benefits, since the forest department may revoke the agreement prior to the harvest.

Some state resolutions note that FPCs will not have a vested interest in the timber until they have protected the forest for at least five years, while revenue sharing will not take place until it is harvested after at least ten years. While such guidelines do provide short-term parameters, they do not reflect tenurial security needs for the long-term management of forests. In West Bengal, the government order states that protection agreements are based on the ten- to twelve-year sal pole rotation. Yet, as market and ecological conditions are changing, this form of silvicultural management is now being questioned by both foresters and villagers. No resolution states what the status of community forest management groups might be after the protected forest is felled. Consequently, the expiration of agreements may create new uncertainties about management responsibilities, providing opportunities for uncontrolled exploitation. It may be more useful to establish automatic extensions for management agreements (roll-overs), provided joint activities are being carried out satisfactorily.

Apart from providing tenurial security through specific clauses in state JFM resolutions, there is also a need to inform management groups of their tenurial rights and formalize this through counter-signed agreements, certificates of usufruct rights, and symbols of authority. When the authority of community management groups is questioned by outsiders or when the group must challenge offenders, such documents are important in demonstrating the legitimacy of the group. Most state forest departments have been slow in translating JFM resolutions into local languages, and in developing effective extension programmes to educate community management groups concerning their rights and responsibilities. Training and extension remain pressing needs, both within the forest department and at the community level.

Joint Forest Use Systems and Microplans

Equitable agreements between forest departments and communities need to be based on jointly developed management plans. Ideally, these should include co-ordinating the use of larger forest tracts including the beat, range, division, and circle. While some resolutions mention community input to microplans, forest department working plans currently are not subject to community input, creating an unresolved tension between the two, especially if they differ substantially. Processes need to be developed to ensure that forest department and management group plans for areas under joint management are compatible and satisfactory for both parties.

Microplans need to be based on sustainable human ecological relationships that emphasize enhanced forest productivity for local residents. In the past, most working plans have emphasized optimizing timber yields. Under JFM agreements, the forest department needs to be able to compromise these objectives to respond to community requirements. In some cases, this will require re-examining old conflicts. For example, if properly managed, long-term rotational agriculture (jhum) can be sustainable and integrated with good forest management. Similarly, if restrictions in space and time are effectively implemented by community managers, grazing on forest lands can be conducted without excessive disturbance. Both Chapters 4 and 5 document that attempts by the forest department to forbid traditional forest use practices, whether migratory grazing, or jhum, have created conflicts with communities and failed to lead to better management.

Community groups need greater opportunities in defining how forest lands can best be sustained productively. Simple research and monitoring could help both parties to assess what practices are viable. Developing good management plans and practices will therefore require the flexibility to experiment.

Many JFM resolutions stress the need for preparing micro-plans with participating communities. Unfortunately, experiences with developing detailed micro-plans have been discouraging. Where hundreds or even thousands of FPCs exist, forestry field staff simply do not have the time or training to prepare in-depth plans with community members. With little training or guidance, and under pressure to prepare many micro-plans, forest range and beat officers may do so hurriedly, resulting in little responsiveness to local priorities and opportunities. Most micro-plans are designed to identify technical and development-oriented investments such as tubewells, roads, and soil conservation structures, rather than to examine the institutional and simple silvicultural requirements. Further, knowledge regarding technical opportunities or options for managing natural forest ecosystems for varied objectives and multiple products is limited. Investments in research, design, and capacity building will be essential to make microplanning a realistic possibility. Even with such a commitment, which currently does not exist, it will be a number of years before forest departments are able to generate substantive working-plans that are contextually grounded in community needs and opportunities. At this early phase in the emergence of community forest management partnerships, there is an urgent need to determine where villages are already protecting forests and to identify high priority areas for programme expansion. It is likely that the first need forest departments face in planning and monitoring the expansion of JFM programmes is a quick method to identify the existing or potential locations of forest protection groups, as well as the condition of local forests.

Revenues, Rates, and Community Remuneration

Most state JFM resolutions provide partial or total rights over subsistence non-timber forest products to the managing community; however, they also clearly indicate the intention of the state to continue to take a substantial portion of major commercial products generated. Even in cases where forest departments are less concerned about maximizing their share of benefits, state revenue departments are concerned over potential losses of income. Given that most of the forest lands targeted for community protection and natural regeneration are highly degraded, villages that do succeed in re-establishing their productivity deserve substantial rewards. This seems particularly just, since forest villages are among the poorest in the country.

Forest departments have often been much kinder to politically well-connected industrial interests in subsidizing prices. As Chapter 6 indicates, foresters and social scientists pioneering JFM in Haryana had to struggle to prevent excessive increases in fibre grass-harvesting rates for poor rope-making communities, despite the fall in grass productivity. Forest departments and government officials need to recognize that it is the poorest segments of Indian society which are attempting to protect a badly abused resource and therefore treat them fairly, not look upon them as a source of income.

In any partnership arrangement of the type suggested by the national and state resolutions, it is important that both the forest department and the participating communities feel that their efforts receive 'fair' remuneration. Fair is not an absolute but would generally be relative to the amount of labour and capital invested by each party, reflecting opportunity costs. Most of the current resolutions cite a specific proportion of the net revenue generated by timber sales. Yet, in West Bengal sample surveys indicate that many villagers are unaware of their share of the timber harvest. The setting of a fixed rate for produce-sharing presents problems. Factors to be considered include the ratio of forest area protected to the number of participating families, the biological productivity of the forest, and the amount of labour required to protect and manage a given tract of forest. In West Bengal, the ratio of forest area to families may vary from 5 hectares per household at one extreme to 0.10 hectare per household at the other. For most FPCs the ratio is between 0.5 to 1 hectare per family. Where the ratio is high — 2 or more hectares per family — forest income is generally substantial, but in areas where the ratio is 0.5 or less, it may be insufficient to meet the opportunity costs of the community management group.

Management costs will also vary between communities and

regions. Communities with fewer families or larger forest tracts may need to spend more time protecting regenerating forests. Communities who were economically dependent on commercial fuelwood-collection but ceased their operations to enter into a joint management agreement with the forest department suffer a greater opportunity cost than those who were only involved in subsistence fuelwood collection. In areas with high grazing or cutting pressures from outsiders, villagers may have to spend much more time patrolling the area than in the forest areas where such pressures are fewer.

For the reasons outlined here, in terms of equity, setting a state-wide fixed rate for sharing timber revenues presents serious problems, particularly when it is presented as a proportion of the net profit with only the costs of the forest departments deducted. Most forest departments have inadequate information regarding the costs of protection to the community. Forest departments offering communities a 25 per cent share of the net timber profit deduct only their own costs from the gross; consequently, community income does not reflect direct or indirect management costs to the villagers. Further, any fixed percentage — as a flat rate — fails to respond to different site characteristics that effect forest income generation. Since participating communities rarely have any knowledge or control over forest department management-costs, they cannot anticipate what income they might derive from management partnership. In the case of the Arabari forest area, one of the first jointly managed areas in West Bengal where sal pole-production sharing was carried out, forest department management costs totalled 53 per cent of the gross. Since the community only received 25 per cent of the net, its share was less than 12 per cent of the gross. One analysis indicated that the actual income of participating FPCs was closer to 6.25 per cent of the gross.[24] Unless there is a process to assess and monitor both partners' costs effectively, the deduction of costs will most likely be unfair and a cause for conflict, arguing for a division based on gross. Ultimately, communities need to feel that they are benefiting fairly from their role in the management partnership. If not, they may lose interest and abandon the programme.

The level of biological productivity is also an important consideration in framing policies regarding produce-sharing. In semi-arid western India, tree growth and biomass production will be slower than in high rainfall areas. Some disturbed forest lands in south-west Bengal still possess healthy sal stumps and other root systems that rapidly regenerate secondary forest growth through coppicing. Within a few years, a community may possess a substantial secondary forest with multiple products being generated. In other forest areas, where stumps have been removed for fuelwood and soil erosion has been extensive, flows of forest products will be considerably slower. If community management costs are to be met, the forest department will need to invest in more capital-intensive enrichment planting and possibly provide additional employment opportunities. To date, JFM policy documents do not address the need for flexibility in ensuring an equitable flow of benefits to participating community management groups operating in different ecological contexts.

Conclusion

For decades, forest departments have acted as sole custodians of vast territories; in essence as 'state zamindars.' However, recent resolutions may represent the beginning of a legislative process that devolves authority for managing forests back to the communities. To succeed, current guidelines need to be strengthened in many areas. There is a need to clarify the relationship between community forest management groups and local governance bodies to ensure that informal user organizations will have the freedom and flexibility to structure themselves and operate effectively in performing critical management tasks. Forest department resolutions need to clarify processes that legitimize community groups without attempting to dominate them. While simply legislating women's participation in forest management is bound to fail, it is a critical element and requires a strong government commitment, both within forest departments and at the community level. Stronger tenurial rights and guaranteed usufructs need clear articulation in all resolutions. Mandated planning systems should require full community participation, benefiting from local knowledge and strengthened by an integration of community goals. Resolutions will need to find ways to deal with conflicts between old working-plan systems, earlier nistar land settlement Act rights and the objectives and needs of emerging community managers. Finally, forest department revenue generation goals, rate-setting procedures, and benefit-sharing arrangements set forth in resolutions should reflect both the significant voluntary labour contributions of the community, as well as the economic hardships such groups face. Most important, resolutions should be designed to generously enable and reward forest users who take on the immense task of rehabilitating India's degraded forests.

Resolutions alone may have little or no effect on field realities and operational activities. Only if they are effectively communicated to forest department staff and village families can these new resolutions help to clarify programme objectives and procedures. These resolutions are currently being translated into local languages in a number of states and circulated to all participants. Meetings are beginning to be organized with forestry staff and participating communities to explain their content and discuss their implications. Through diagnostic research, programme monitoring, and open discussion with participating groups, new ideas are emerging and a number of states are revising their JFM resolutions. These ideas, in turn, are strengthening policy instruments and the programmes they support, facilitating community involvement in managing India's natural forests. To the extent that policies and programme activities can respond in a supportive manner to the problem-solving strategies being developed by communities and foresters, JFM offers a promising opportunity to respond to India's forest management problems in a socially and ecologically sound manner.

Notes

1. W.R. Bentley, G.B. Singh, and N. Chatterjee, 'Tenure and Agroforestry Potentials in India' (paper presented at the International Workshop on Tenure Issues in Agroforestry, New Delhi, 1988).

2. As long as 5000 years ago, Vedic records indicate that an extensive body of information existed regarding the uses of various woods, medicinal plants, botany, and physiology. See Anil Rawat, 'Life, Forests, and Plant Sciences in Ancient India', in Ajay S. Rawat (ed.), *History of Forestry in India* (New Delhi: Indus Publishing, 1991), pp. 241–64.

3. G.S. Puri, V.M. Meher-Homji, R.K. Gupta, and S. Puri, *Forest Ecology*, vol. 1 (New Delhi: Oxford and IBH, 1990), p. 4.

4. U. Rane, 'The Zudpi Factor', *Sanctuary* 7:324, as cited in Madhav Gadgil,

'Deforestation: Problems and Prospects', in Ajay S. Rawat (ed.), *History of Forestry in India* (New Delhi: Indus Publishing, 1991), p. 13.
5. Mark Poffenberger (ed.), *Keepers of the Forest* (New Hartford, Conn.: Kumarian Press, 1990), p. 12.
6. Richard P. Tucker, 'The British Empire and India's Forest Resources: The Timberlands of Assam and Kumaon, 1914–1950', in John F. Richard and Richard P. Tucker (eds), *World Deforestation in the Twentieth Century* (Durham, N.C.: Duke University Press, 1988), p. 91.
7. Madhav Gadgil, 'Deforestation: Problems and Prospects', Founder's Day Lecture, Society for the Promotion of Wastelands Development (New Delhi, 12 May 1989), p. 16.
8. Section 2, Forest Act of 1865, as cited in Jonathan M. Lindsay, 'Law and Community in the Management of India's State Forests' (draft manuscript, 29 July 1993), p. 38.
9. The authors acknowledge both Ramachandra Guha's and Jonathan M. Lindsay's thoughtful research in reconstructing the debate of the mid-nineteenth century that continues even today.
10. See J.M. Lindsay, 'Law and Community in the Management of India's State Forests', p. 55.
11. Ibid., p. 52.
12. Ibid., p. 58.
13. See W.R. Bentley, G.B. Singh and N. Chatterjee, 'Tenure and Agroforestry Potentials in India', p. 231.
14. Madhav Gadgil and Ramachandra Guha, *Fissured Land* (New Delhi: Oxford University Press, 1991).
15. Tom Kenworthy, 'Forest Service Chiefs Wink at Timber Theft, Hill Told', *Washington Post,* 7 October 1993 (The Federal Page).
16. Section 6, Forest Policy of 1952, as cited in Lindsay, p. 84.
17. Ibid.
18. Ibid., p. 121.
19. Government of Orissa, Resolution no. 10F (Pron)–47–88/23638/FFAH, Protection of Reserve Forest Areas by the Community, 13 October 1988.
20. Neera and Kundan Singh, 'Case Report on the Bruksho O' Jeebaro Bandhu Parishad' (Bhubaneshwar: Vasundara, n.d.), pp. 18–19.
21. Personal communication from Neera and Kundun Singh.
22. Rekha Singhal, 'Gender Issues in Joint Forest Management: More Than a Trendy Issue' (Bhopal: IIFM, 1994), p. 25.
23. Personal communication from Neera Singh, May 1993.
24. Personal communication cited in Lindsay, p. 136.

Table 2.1 Recent government guidelines for joint forest management*

	Orissa	West Bengal	Bihar	Gujarat	Rajasthan
Date of issue of resolution	14–12–88	12–7–89	8–11–90	13–3–91	16–3–91
Forest category	Reserve	Degraded	Degraded (protected)	Degraded	Degraded
Participants	Adjoining villagers	Economically backward people	1 person from each family	Persons interested in forest development	Willing villagers
Management unit	1 forest compartment	Forest beat	Village	Village	Maximum 50 ha. 1 village, if possible
Executive committee: People's representative	3 or less	6 or less	Dependent on forest	Minimum 2 women, any other	–

Forest dept. representative	Forester	Beat officer	Vanpal	—	According to state government rules
Share of members:					
A. Non-timber forest products (NTFPs)	—	Cashew, 25%, *Tendu* sal seed, leaves on approved tariff, rest free	Dry branches, grass, leaves free; other produce available at market price	Dry branches and NTFP free	Grass and fodder (after 5 years) free; NTFP (except bamboo) collection according to provisions of management plan
B. Timber	All bonafide subsistence needs of timber and fuelwood free	25% of net income except in certain areas	1/3 share of income deposited as village development fund	If state financed, 25%; otherwise, 80%	60% of net income after deducting all government expenditures

(cont.)

Table 2.1 (cont.)

	Madhya Pradesh	Tripura	Maharashtra	J & K
Date of issue of resolution	10–12–91	20–12–91	16–3–92	19–3–92
Forest category	Sensitive to damage and degraded	Degraded	Degraded and barren lands	Degraded
Participants	Willing villagers	Families with at least 1 wage earner	Panchayat and FPC consisting of all villagers	1 person from each family of adjoining villages
Management unit	1 village	Natural regeneration 500 ha. plantation 300 ha.	Not defined	Not defined
Executive committee:				
People's representative	5 or more	5 or less	6 (2 women and 2 SC/ST**)	11 (2 women and 2 SC/ST)
Forest dept. representative	Ranger	Beat officer	Forester	Ranger

Share of members:

A. *Non-timber forest products (NTFPs)*	All forest produce free and 30% of net income from nationalized NTFPs	All NTFPs free	All NTFPs except cashew and *Tendu* free	All forest produce free
B. *Timber*	30% of net revenue; 20% for damage-sensitive areas	All bonafide needs met and 50% of net surplus revenue	Different methods of distribution in different areas such as block plantation scheme, nistar, etc.	25% of net revenue from final harvest in cash/kind

* Adapted from Table 2 in 'Joint Forest Management: Concepts and Opportunities'. Proceedings of the National Workshop at Surajkund, August 1992. Society for the Promotion of Wastelands Development, New Delhi, pp. 8–9.

** SC/ST = Scheduled Caste/Scheduled Tribe

Chapter 3

Dietrich Brandis and Indian Forestry: A Vision Revisited and Reaffirmed

Ramachandra Guha

In recent years there has been a growing movement for democratization of forest management in India, for widening the
sphere of popular participation by making forest managers more
responsive to the needs of local communities. Grassroots organizations active in forest areas have repeatedly urged a complete
overhaul of forest management. State policies, they contend,
have excluded the dominant majority of the Indian population
from the benefits of forest-working while favouring the interests
of a select group of industries and urban consumers. At the
same time, they have called for a reorientation of forest policy,
towards serving the local subsistence sector more directly. The
forest department has been asked to shed its reliance on punitive
methods and to adopt a more participatory and community-
oriented system of management.[1]

Keeping this debate in mind, this essay deals with a curious
and little known paradox. It argues that the founder of the Indian
Forest Service, Dietrich Brandis, had anticipated that a system of
forestry founded exclusively on state control would lead to great
disaffection in the countryside. Keenly cognizant of village rights
and demands, Brandis had himself suggested a system of forest
management based on a collaborative relationship between the
state and local communities. Ultimately the Brandis route was not
taken by Indian forestry, which preferred instead to uphold the
principle of state monopoly over forest land. Yet his ideas,
elaborated in this essay, have a strong contemporary resonance.
While terms like 'social forestry,' 'community forestry,' and 'joint
forest management' are of recent provenance, the principles they

embody would have been readily recognized and perhaps warmly commended by the first Inspector General of India's Forest Department.

Brandis before India

Dietrich Brandis was born in Bonn, Germany, on 31 March 1824, to a highly educated family of professionals. His paternal grand-father had served as personal physician to the King of Denmark, and his father as an educational adviser to the King of Greece. Educated at the universities of Bonn and Gottingen, Brandis was trained in botany and chemistry. Joining the faculty of Bonn University in 1849, having already published a series of research papers and discovered two rare plants, Brandis appeared well-positioned for a career in botanical teaching and research.[2]

Brandis was drawn to India by his marriage to the English widow of a Danish doctor who had served at Serampore, in Bengal. Curious to continue his botanical studies in an unexplored region of India, Brandis took advantage of his wife's connections to obtain the post of superintendent of the teak forests of Pegu, British Burma, in January 1856. Having been granted a two-year leave from Bonn University, the botanist's primary impulse to explore an outlying province of the British Indian Empire was a scientific one. Never returning to his teaching post, Brandis proceeded to stay twenty-seven years in India.[3]

The rich teak forests of Burma were a valuable resource for the British crown, providing timber of exceptional quality for shipbuilding. But the forests were being recklessly worked by private entrepreneurs (mostly European), and it was Brandis's task to realign their management on a sound ecological footing. On the political front, Brandis was forced to fight a long and difficult battle to keep private timber merchants away from the forests, which he sought to bring increasingly under state control. Silviculturally, Brandis focused on teak-dominant forest tracts, seeking through protection and enrichment plantation to enhance the proportion of this most highly valued species in the standing forest. At the same time, Brandis was successful in augmenting state revenues from the forest, a noteworthy achievement for a commercially minded colonial government.[4]

Brandis's work in Burma convinced him of the vital importance of state control over forest ownership and management. The fusion of forestry with the state and its logical corollary, and the hostility to private enterprise were among the basic tenets of scientific conservation to which Brandis wholeheartedly subscribed.[5] Where Brandis was to depart radically from the consensus among 'scientific' foresters was in his enthusiasm for community ownership as a parallel but complementary system to state forestry. With valuable teak forests located in the territory controlled by militant tribes such as the Karens, Brandis was only too aware of the potentially disruptive effects of a state forestry system that was too extensive and exclusionary. As he recalled later, his Burmese experience forewarned him of the need to 'conciliate the people in the vicinity of the forests . . . '[6] During his tenure in Burma, Brandis helped in developing a unique method of conciliation and village participation known as the *taungya* system. Communities of shifting cultivators were encouraged to grow food crops in the forest, provided they planted teak saplings between rows of rice.[7]

Brandis as Early Social Forester: The Mysore Controversy

In recognition of his outstanding work in Burma, Brandis was transferred to wider duties in the government of India in December 1862. The enormous timber demands of the railway network had sensitized the British in India to the importance of systematic forest management. At this time, the British were the world leaders in deforestation, having devastated their own resources as well as the forests of Ireland, southern Africa, and north-eastern United States to obtain timber for shipbuilding, iron smelting, and for agricultural conversion.[8] In contrast, Germany was the leading European nation in forest management.

During his nineteen years as Inspector General of Forests (IGF), Brandis laid the foundations that underpinned state forestry in India. A man of great energy, he toured widely on the subcontinent, writing authoritative reports on recommended directions for forest management in the different provinces of British India.[9] In the realm of silviculture, he formulated the systems of valuation and forest-working still widely in use today. True to his university background, Brandis also emphasized research and training, establishing a forest college to train lower staff, arranging overseas training for higher officials, and setting up the Forest Research Institute at Dehradun.

While an assessment of the scientific and administrative aspects of Brandis's legacy would be beyond the scope of this discussion, his *sociology* of forest management, namely his understanding of the social contexts of state forestry operations in India, is of noteworthy interest. Brandis's views are immediately distinguishable from those of his epigones who have since headed the Indian forest establishment. Retrospective accounts of Indian forestry — most prominently E.P. Stebbing's official history — juxtapose 'scientific' forestry under state auspices with the customary usage of forests by rural communities. Such accounts of community practices have been typically referred to as erratic, unsystematic, and unmindful of long-term sustainability — charges that Brandis strongly disagreed with. Although faced with a barrage of criticism, contemporary forest officials have likewise justified their territorial control over 23 per cent of India's landmass on the grounds that they alone possess the technical skills and administrative competence to manage public forest lands.[10]

Clearly, Dietrich Brandis shared a larger professional faith in the scientific status of sustained-yield forestry and the inevitability of state control. His divergence arose from his colleagues' skepticism of the knowledge base of rural communities. Contrary to views held by departmental historians such as Stebbing, state forestry for Brandis did not operate on a clean slate; rather, it was an intervention in an already existing local system of control and use.

As an example, Brandis wrote with admiration about the widespread network of sacred groves in India. At various times, he referred to these as the 'traditional system of forest preservation' and 'illustrations of indigenous Indian forestry'. He had himself discovered sacred woodlands to be 'most carefully protected' in many districts and in nearly all provinces of British India — from the Devara Kadus of Coorg in southern India to the sacred groves of Mewar in the north-west. While the prevalence of sacred groves provided evidence of a decentralized system of forestry, at the other end of the spectrum Brandis also documented forest reserves well-managed by Indian rulers. He was particularly impressed by the chiefs of Rajasthan (e.g., the princes of Bharatpur, Jaipur, Udaipur, and Kishengarh), whose hunting preserves provided game for the nobility as well as a permanent supply of fodder and small timber for the peasantry. In their strenuous efforts to preserve dry thorn and scrub forest in an arid climate, these Rajput rulers set a good example according to Brandis, one 'which the forest officers of the British government would do well to emulate.'[11]

Brandis's faith in the role and capacity of local institutions in forest management was perhaps most clearly evident in his larger vision of Indian forestry. Even as he was consolidating control of the state over valuable forest areas, he was also proposing the creation of an extensive, parallel system of communal forests for village use. A series of reports and memoranda illustrate how the IGF tried for more than a decade to persuade the colonial government that a strong system of village forests was indispensable to the long-term success of state forestry.

The first major effort by Brandis in this regard can be dated to 1868, only four years after the creation of the Indian Forest Service as a formal institution. In that year, Brandis visited the southern province of Mysore — then under British rule, but to revert in 1881 to its Hindu ruler — and wrote a persuasive report recommending the constitution of village forests in that territory. Noting that waste and uncultivable lands comprised more than 50 per cent of the village area, he offered a detailed scheme suggesting that this degraded land could provide the basis for a system of village forests in at least 52 out of the 83 *talukas* (subdistricts) of Mysore. Crucial to the success of Brandis's scheme was the inclusion of grazing lands in village forest tracts. Each proposed forest-pasture unit would be of sufficient area for management on a rotational coppicing system, with freshly cut areas closed to grazing and fire. No portion of the village forest was to be demarcated for cultivation without the sanction of the commissioner. Ideally, each hamlet would have its own forest, but in many cases it would be necessary to constitute a tract for collective use by a group of villages. In the scheme, nine commercially valued species, including sandalwood and teak, were to be reserved for the state.

Brandis envisioned that such forests would provide numerous items *gratis* to the villagers: firewood for domestic consumption, and for sale by poor headloaders; wood for agricultural implements, cart construction, and repair; wood, bamboo, and grass for thatching, flooring, and fencing; leaves and grass for green manure; and pasture fodder, excluding those blocks closed under the recommended rotational system. Upon payment of a nominal use-fee, wood would also be available for house construction and raw material for local artisans. In the IGF's scheme, these community forests would be placed under a parallel administrative system, with a designated village forester for each unit, a forest ranger overseeing all the village forests in each taluka, and a head forest ranger for the district as a whole who would report to an assistant conservator of forests. Brandis anticipated a self-supporting system; revenues from the village forests would cover administrative costs, and any surplus would be reinvested in local improvements. With this self-reliant approach, peasants would gradually acquire an interest in the maintenance and improvement of their forests.[12]

Forwarding his Mysore report to the government of India, Brandis noted that it was the first of a series of measures he would propose in various provinces for the improved management of extensive wastelands, which would be excluded from the state forest-system. These measures would serve as a prelude to recommendations for a countrywide system of community forests.[13] Meanwhile, to the commissioner of Mysore, Brandis grounded his proposal in the context of agricultural improvement in general, noting that ' . . . the art of agriculture in Mysore is further advanced than in many other parts of India; great attention is paid to the supply of water, to manuring, to the working of the ground, and to some extent also, to a rotation of the crops; it does not therefore seem unreasonable to expect, that the *ryots* [peasants] may be brought to advance another step, and in the management of their grazing grounds and jungle lands, to adopt a regular system of rotation, which with a few other measures, will do much to make those lands more productive.'[14]

Circulated for comments among district officers in Mysore, the majority of the opinions solicited on Brandis's report supported the creation of village forests. A lone Indian official even argued that a separate state establishment was not required, for the peasants could manage the forests themselves. The deputy superintendent of Bangalore, a man with considerable knowledge of local botany, geology, and culture, also added breadth to Brandis's ideas by suggesting that a council of men (one elected from each village) form a forest court to settle disputes arising out of the management of village forests.[15]

Predictably, other officials strongly dissented with the controversial Mysore report, arguing that the proposed scheme would lead to a loss of state revenues while undermining the powers of district officials. They argued that open-access, communal resources would suffer from short-term individualistic exploitation. The village communities of Mysore, without cohesion due to caste factions, 'could not be entrusted with the powers or competence to perform the functions assigned to them in [Dr. Brandis's] scheme.' Others confirmed the notion that the scheme would fail, 'as each man, when the least removed from supervision, would cut whatever he might require for himself without any regard to the interests of his neighbours . . . '[16]

The chief commissioner of Mysore opted for the latter pessimistic viewpoint. In lieu of forests controlled by villagers, he recommended that unreserved forests remain in the control of revenue officials, that convenient depots be operated by the state for the sale of forest produce, and that individual ryots be encouraged to plant trees on their private lands. In a classically colonial attitude, the government of India concurred, claiming that 'the prejudices and rivalries of Natives might be excited if men of different classes and castes shared in the same forests.'[17]

This judgment was not accepted lightly by Brandis. In defiance, he reviewed the case afresh and offered another forceful plea in favour of village forests. The constitution of such forests in Mysore, he suggested, could provide an indirect means for the British to ensure agricultural output and prosperity once the territory reverted to its Wodeyar rulers. He also conceded to the Mysore officials, with views even more advanced than his, that forest department control could give way in due course, with the 'leading men' of each village assuming responsibility for the management of the designated areas. Brandis drew pointed attention to the flourishing system of community forests on the continent, where scientific foresters exercised technical supervision over woodland managed for the exclusive benefit of villages and small towns. About the European model Brandis wrote,

'Such Communal Forests are a source of wealth to many towns and villages in Italy, France and Germany; property of this nature maintains a healthy spirit of independence among agricultural communities; it enables them to build roads, churches, school-houses, and to do much for promoting the welfare of the inhabitants; the advantages of encouraging the growth, and insisting on the good management of landed communal property, are manifold, and would be found as important in many parts of India as they have been found in Europe.'[18]

However, this reference to European parallels failed to move Brandis's plan off the shelf. Undaunted, he returned to the subject on a later visit to Mysore, where he still hoped that at some future date he would 'find the advantages of true communal forests recognized not only in Mysore but in all parts of India where a village organization exists,' for it seemed 'particularly desirable to strengthen the old village organization by consolidating and ameliorating the grazing grounds, forests and waste land of the village community.'[19]

The 1878 Indian Forest Act

In many respects, Brandis's controversial Mysore proposals initiated a major debate in the colonial bureaucracy leading to the passing of the 1878 Forest Act. A comprehensive piece of legislation designed to protect and further extend state control over forest areas throughout India, the Act, with minor modifications, is still in operation.[20]

Encompassing the sphere of the controversy, three major schools of thought emerged. The first advocated *total* state control over *all* forest areas. Its foremost proponent, B.H. Baden-Powell, rested his case on a highly selective and legalistic reading of Indian history, wherein 'Oriental Sovereigns' had the right to dispose of forests and wastelands as they wished, a right transferable to their British successor. Baden-Powell's position found wide acceptance among forest officers who justified a massive system of state control as the only check on individual self-interest and short-sightedness.[21]

The directly opposite view, held pre-eminently by the Madras government, totally rejected state intervention in the belief that tribals and peasants should exercise complete control over forest land. One particularly vocal member of the Madras Board of Revenue went so far as to suggest that the state had no right over uncultivated lands that were invariably 'village property, not village privilege'. In this southern presidency 'nearly all the jungles and forests [were] within village boundaries' and the state could at best play a subsidiary role in the management of such forest areas.[22]

Rejecting both these extremes, Brandis advocated a balanced but *restricted* takeover of forests by the state. He justified this middle course both on the grounds of justice, — respect for customary village rights — and efficiency, urging the administration 'to demarcate as state forests large and compact areas of valuable forests as can be obtained free of forest rights of persons', while leaving the residual area, smaller in extent but more conveniently located for their supply, under the control of village communities. He hoped ultimately for the creation of three broad classes of forest property, based on the European experience: state forests, village and other community forests, and private forests. State ownership had to be restricted, argued the IGF, due to the 'small number of experienced and really useful officers' in the colonial forestry service, and out of deference to the needs and wishes of the local population. Brandis stated, 'The trouble of effecting the forest rights and privileges on limited, well-defined areas is temporary and will soon pass away, whereas the annoyance to the inhabitants by the maintenance of restrictions over the whole area of large forest tracts will be permanent, and will increase with the growth of population.'[23]

Once again, Brandis's views did not prevail. The 1878 Forest Act was firmly based on the contrasting proposals of Baden-Powell. In subsequent years, this legislation formed the basis for the designation of huge areas as state forests — now covering over 23 per cent of the country's land area — owned, closely controlled, and regulated by the Forest Service. While the Act did have a provision for the constitution of village forests, the option was not exercised by the government except in a few isolated instances.[24]

Brandis Post-India

In the midst of the Forest Act debate, Brandis's immediate superior, the agricultural secretary, remarked that the IGF's 'views as to rights of aboriginal tribes, forest villages, etc. are to my mind clearly in advance of my own and a fortiori of those of the government of India.'[25] Indeed, these beliefs in the forest rights of communities have remained well in advance of those of successive Indian governments, whether colonial or democratic.

What made Brandis's articulate views so strikingly different from those held by his peers in the colonial system and by forest officials since? While a full explanation may be impossible, first, he originated from a nation that had not yet acquired an overseas empire. Thus, he could believe, unlike Baden-Powell, that the forest rights of Indian villagers were analogous to those of European peasants. At a personal level, there were various indications of Brandis's concern for social reform, which was exceptional for its time and milieu. With a desire for voluntary work developed early from his mother, even in his retirement in Germany he was to work actively with workingmen's clubs.[26] On a conceptual level, Brandis had a far keener appreciation than other colonial officials of how, particularly in the Indian context, the produce of the forests was vital to the ecological stability of the agrarian economy. Finally, Brandis's enthusiasm for community forestry also stemmed from a shrewd awareness of the adverse social consequences — in terms of rural unrest — given an overly exclusive system of state forest authority.

Whether born out of intellectual, tactical, or biographical reasons, Brandis's support for the creation of a nationwide network of village forests was remarkably persistent. Shortly after relinquishing the post of IGF, he likened systematic forestry in India to 'a plant of foreign origin, and the aim must be to naturalize it.' On the social side, this process of indigenization could be accomplished by encouraging native chiefs, large proprietors, and especially village communities to develop and protect forests for their own sustained use. In the end, the initiative lay with the government, and here Brandis emphasized the enormous gains from a successful system of communal forests. 'Not only will these forests yield a permanent supply of wood and fodder to the people without any material expense to the State,' he wrote, 'but if well managed, they will contribute much towards the healthy development of municipal institutions and local self-government.'[27]

Thirteen years later, Brandis returned to the subject in an extraordinary essay recommending the extended employment of

Indians at various levels of the Forest Service. Long after he had severed all formal contacts with British India, his deep concern that Indian forestry cease to have 'the character of an exotic plant, or a foreign artificially fostered institution' continued. This concern was integral to his larger democratic vision for forestry on the subcontinent. With their social as well as silvicultural education in mind, Brandis suggested that 'native' Indian foresters be sent to study the forestry system in Germany, noting that if sent there, foresters 'will find that the villages, which own well managed communal forests, are prosperous, although now and then they complain of the restrictions which a good system of management unavoidably imposes. What Indian Forest Officers will learn in this respect in Germany will be really useful to them in India.'[28]

Brandis may have despaired and given up on British officials in India taking his proposals for the constitution of village forests seriously.[29] The more indirect approach, wherein Indian forest officers trained on the continent, might better expose the benefits of community forests. Nonetheless, Indian officials, whether trained in Germany or not, have been for the most part, hostile to the suggestion that local communities could be encouraged to manage forest areas for their own use. It is this territorial monopoly and indifference to the demands of rural communities that has made the Indian Forest Service the object of such relentless criticism in recent years.

Although slow, significant changes are emerging as a consequence of the ongoing debate on forest management in India. In rural villages throughout the country, non-governmental organizations have worked diligently to foster a sense of community ownership in the small areas of forest and pasture land still outside the purview of state control. In the vast domain of government forests, meanwhile, radical innovations have been attempted in the state of West Bengal and others, namely, the creation of thousands of village Forest Protection Committees working on state land, managed under a partnership arrangement between the forest department and the surrounding villages. Steps are currently being taken to expand and extend the experiment to other forested regions of India.[30]

Without their knowledge, advocates of joint forest management (JFM) are, in effect, reviving and reaffirming Dietrich Brandis's vision for Indian forestry. Striking parallels can be seen between the ideas of Brandis and the ideas underlying the successful application of JFM in West Bengal. With respect to the role of forest-dependent communities, for example, there is a shared faith in indigenous knowledge, in the management capacity and robustness of local institutions, and above all, a sharp focus on local access to the usufruct of forests. Again, with respect to the role of the state, there is a common recognition of the essentially advisory role of the forest department, of its need to collaborate with rather than strictly regulate customary use, and of the justice of sharing revenues from forest-working with the villagers. Finally, both Brandis and JFM proponents seem to converge in their larger vision for forest policy in India, a vision which in my understanding consists of three central principles: (1) that benefit sharing — between state and community — and local control are to be the key incentives to ensure sustainable management and to minimize conflict; (2) that community-controlled forests, under a JFM kind of system, would work as a complement to a network of more strictly protected areas, farther from habitations that continue under direct state control; and (3) that the restricting of state control to these latter areas is vital on grounds of equity (i.e. the respect for local rights and demands), efficiency (i.e. as the most feasible course, with the state not biting off more than it could chew), and stability (i.e. as the most likely way to lessen conflict).

It has been argued that JFM, as an attempt to reverse or mitigate state monopoly over forest ownership, protection, and management, constitutes a significant departure from past trends. However, the evidence presented in this essay might suggest that contemporary attempts at fostering community forestry and JFM are a reaffirmation of the ideas put forward by the founder of the Indian Forest Department. Thus in the late twentieth century, as in the late nineteenth century, a movement has arisen for the creation of a democratic and participatory system of forest management, one founded not on mutual antagonism but on a genuine partnership between the state and the local communities. Dietrich Brandis's vision for Indian forestry, so abruptly cast aside in the 1860s and 1870s, may yet come to prevail.

Notes

1. See, for example, People's Union for Democratic Rights, *Undeclared Civil War* (New Delhi: PUDR, 1982); Walter Fernandes and Sharad Kulkarni (eds), *Towards a New Forest Policy* (New Delhi: Indian Social Institute, 1983); Ramachandra Guha, *The Unquiet Woods: Ecological Change and Peasant Resistance in the Himalaya* (New Delhi: Oxford University Press and Berkeley: University of California Press, 1989).

2. These biographical details are drawn from Herbert Hessmer, *Leben und Werk von Dietrich Brandis: 1824–1907* (Dusseldorf: Westdeutscher Verlag, 1975). I am grateful to Professor S.R.D. Guha for translating relevant portions of the book for my use.

3. Ibid.

4. Anonymous, 'Dietrich Brandis: The Founder of Forestry in India', *Indian Forester* 10(8): 347–57, 8 August 1884.

5. Samuel Hays, *Conservation and the Gospel of Efficiency: The Progressive Conservation Movement, 1880–1920* (Cambridge, Mass.: Harvard University Press, 1958).

6. Note by Dietrich Brandis dated 18 August 1870, in Revenue and Agriculture (Forests) [hereafter Forests], B Prog. no. 82, August 1870, National Archives of India, New Delhi [hereafter NAI].

7. H.R. Blanford, 'Regeneration with the Assistance of Taungya in Burma', *Indian Forest Records*, vol. 11, pt. 3, 1925.

8. The massive deforestation in India undertaken by the British in the first century of their rule is described in, among other works, H. Cleghorn's, *The Forests and Gardens of South India* (London: W.H. Allen, 1860), and E.A. Smythies', *India's Forest Wealth* (London: Humphrey Milford, 1925). For a classic account of the links between British overseas expansion and deforestation; see R.G. Albion, *Forests and Sea Power* (Cambridge, Mass.: Harvard University Press, 1926).

9. For example, D. Brandis, *Suggestions Regarding Forest Management in the Northwestern Provinces and Oudh* (Calcutta: Government Press, 1881), and *Suggestions Regarding Forest Administration in the Madras Presidency* (Madras: Government Press, 1883).

10. E.P. Stebbing, *The Forests of India*, 3 vols (London: John Lane, 1922–27); J.B. Lal, *India's Forests: Myths and Reality* (Dehradun: Natraj Publishers, 1990); cf. also the interesting discussion in Louise Fortmann and Sally Fairfax, 'American Forestry Professionalism and the Third World — Some Preliminary Observations', *Economic and Political Weekly*, 12 August 1989.

11. D. Brandis, *Indian Forestry* (Working: Oriental Institute, 1897), pp. 12–14; idem, *The Distribution of Forests in India* (Edinburgh: Mcfarlane and Erskine, 1873), pp. 24–5; Brandis to James Brown, dated 15 August 1881, in Forests, B Prog. nos 30–32, September 1881, NAI. The IGF also displayed an early ethnobotanical interest in indigenous tree and plant classification, in circulating a list of local names and urging 'younger

officers, with more leisure and more extensive opportunities, to take up the study of the names of trees and shrubs used by the [tribes] of Central India.' See Brandis, *Suggestions Regarding Forest Administration in the Central Provinces* (Calcutta: Government Press, 1876), appendix.

12. D. Brandis, 'On the Formation of Village Forests in Mysore' (report dated May 1868), in 'Forests', Prog. nos. 63–66, June 1870, NAI. All archival sources in the rest of this section, unless otherwise mentioned, are from this source.

13. Brandis to the Secretary, Public Works Department, dated 9 January 1869.

14. Brandis to the Secretary of the Commissioner of Mysore, dated 22 May 1868.

15. See B. Krishnaiengar, deputy superintendent, Kolar, to superintendent, Nandidroog Division, dated 14 January 1869; J. Puckle, deputy superintendent, Bangalore, to superintendent, Ashtagram Division, dated 16 February 1869.

16. L. Ricketts, deputy superintendent, Mysore, to superintendent, Ashtagram Division, dated 2 February 1869; Inspector General Cumming, deputy superintendent, Shimoga, to superintendent, Nagar Division, dated 2 February 1869.

17. Chief commissioner of Mysore to the secretary of the Public Works Department, government of India, dated 6 May 1869; Resolution no. 172 in the 'Foreign Department', dated 24 June 1870.

18. Note on 'Village Forests, Mysore' by D. Brandis dated 28 May 1870.

19. 'Memorandum by D. Brandis, IGF, on the district scheme of Mysore,' dated 1 June 1874, in Forests, B Prog. nos 12–15, June 1874, NAI.

20. This section draws largely on Ramachandra Guha's, 'An Early Environmental Debate: The Making of the 1878 Forest Act', *Indian Economic and Social History Review* 27(1), 1990. This essay may be consulted for more details and sources.

21. See especially B.H. Baden-Powell, 'On the Defects of the Existing Forest Law (Act XIII of 1865) and Proposals for a New Forest Act', in B.H. Baden-Powell and J.S. Gamble (eds.), *Report of the Proceedings of the Forest Conference, 1873–74* (Calcutta: Government Press, 1875).

22. See especially Forests, Prog. nos 1–52, March 1879, NAI; and Legislative Department, A Prog. nos 43–142, March 1878, NAI.

23. See especially D. Brandis, 'Explanatory Memorandum on the Draft Forest Bill,' dated 3 August 1869, in Forests, B Prog. nos 37–47, December 1875, NAI; idem, *Memorandum on the Forest Legislation Proposed for British India (Other Than the Presidencies of Madras and Bombay)* (Simla: Government Press, 1875).

24. See B. Ribbentrop, *Forestry in British India* (Calcutta: Government Press, 1900).

25. Note by A.O. Hume, dated 19 August 1874, in Forests, Prog. nos 43–55, March 1875, NAI. Hume, in 1885, helped set up the Indian National Congress.

26. See Herbert Hessmer, *Leben und Werk von Dietrich Brandis: 1824–1907*, pp. 6, 392.
27. D. Brandis, 'The Progress of Forestry in India', *Indian Forester* 10(11): 508–10, November 1884.
28. D. Brandis, 'Indian Forestry: The Extended Employment of Natives', *The Imperial and Asiatic Quarterly Review*, April 1897, especially pp. 255–6.
29. Interestingly, an agricultural chemist visiting India some years previously, had sharply criticized the forest department for its narrow commercial orientation, recommending instead the creation of 'Fuel and Fodder Reserves' to more directly serve the rural population. However, his suggestions fell on deaf ears. See J.A. Voelcker, *Report on Indian Agriculture* (Calcutta: Government Press, 1893).
30. K.C. Malhotra and Mark Poffenberger (eds), *Forest Regeneration Through Community Protection: The West Bengal Experience* (Calcutta: West Bengal Forest Department, 1989); Madhav Gadgil and Ramachandra Guha, *Ecology and Equity: Steps Towards an Economy of Permanence* (Geneva: United Nations Research Institute for Social Development, 1994), Chap. 7.

Chapter 4

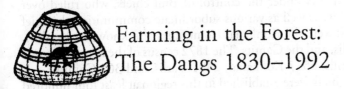

Farming in the Forest:
The Dangs 1830–1992

David Hardiman

In early 1990, a group of *adivasis* who lived in the Dangs District of Gujarat entered an area of reserved forest and began clearing it for cultivation. They claimed that the area was *juni gaothan*, meaning the site of an old village in which they had an ancient right of cultivation. The forest department of the Gujarat Government did not accept this claim and arrested about ninety people for violating the forest laws. Rather than crush the movement, as anticipated, the authorities met with an unexpected resistance. The people, they found, had a fervent belief in the righteousness of their act.[1]

In this struggle, which grew in strength in the following two years, two systems of belief came into sharp conflict. On the one hand, there were the adivasis, who believed that through ancient custom they had the right to cultivate in their homeland when and where they pleased; on the other hand, there were the forest officials who asserted their full right under the law to determine exactly how all tracts of reserved forest should be utilized. It is this conflict, which has continued in the Dangs for well over a century, that forms the subject of this chapter.

Background

The Dangs, a predominantly adivasi region, lies between the fertile plains of south Gujarat and the rugged plateau of western Maharashtra (see Figure 4.1). It is a land of mountain ranges, covered in forest and intersected by deep river valleys. During the medieval period, it was under the control of Bhil chiefs, who ruled over other Bhils as well as various subordinate communities, the chief of which were the Konkanas (sometimes also known as Kunbis), the Varlis, and the Gamits. The 1872 census of the Dangs counted 7426 Bhils, 6517 Konkanas, 2491 Varlis, and 302 Gamits.[2]

Bhil chiefs were established in this region at least four hundred years ago, for the British traveller Sir John Hawkins referred in 1608 to the Dang chief of 'Cruly' — most probably the Bhil chief of Kirli. The Konkanas migrated into the region from the coastal areas of Maharashtra to the south-west — the Konkan. This is borne out not only by their community name, but also their language, which is a dialect of Marathi with certain north Konkani elements. According to an oral tradition, the community migrated north as a result of the terrible Durgadev famine of 1396–1408 that devastated the Konkan region.[3]

Early Resource Use Systems

The Bhil chiefs appear to have been ranked according to the power exercised by any one of them at a particular time. Almost all Bhils were related in some way or the other to a chief. The hierarchy among the Bhils was based on power and status rather than material wealth. Their simple dwellings were built of a wooden frame with bamboo walls smeared with a mixture of cow dung and mud, and the roof thatched with leaves. The chiefs themselves did no field labour, expecting their non-Bhil subjects to grow crops for them. The non-Bhil cultivators were required to pay an annual land tax, the amount of which varied according to whether cultivation was by plough or hand. It was normally paid in grain.

The Konkanas were associated with *rab* cultivation, which was confined to the flatter areas of the valleys. Debris from the forest was piled on a small plot of land and burned. Seeds, normally of *nagli* (a grain) or rice, were sown. Once the monsoon rains came, the area around this plot was either ploughed or worked with hoes. It seems to have been common for Konkanas to use bullock-drawn ploughs.[4] When the seeds had sprouted, they were transplanted to the prepared land. Weeding had to be carried out periodically, and the crop was harvested soon after the monsoon.

The rab of burned forest-waste is a particularly good medium for seed growth, and yields are high. The Dangs, with its abundant supply of such waste, was well-suited for such agriculture. As a rule, an area was cultivated only for two or three years. After this the soil was left to fallow and a fresh patch of land was sought.

The Bhils and the Varlis opted for *kumri* or *dalhi* cultivation, which required no ploughing or weeding.[5] A portion of the forest was selected — often on the top of a plateau or the higher side of a valley — and the forest was cleared of scrub. Once it had dried, it was burned. Seeds — normally of nagli — were sown in the ashes. The seeds sprouted and grew once the rains came, and harvesting was possible by the end of the monsoon. Crop yields were low, as the quality of soil on these plateaus and slopes was poor. It appears to have been a common practice to select a fresh patch of forest each year.[6]

The Bhils and the Konkanas used different techniques and terrains for their agriculture and the two methods complemented each other. Both required plenty of forest and space. Villages were very small, and few and far between. According to an official, writing in 1877: 'The Dangs may be termed a sea of forest and trees, with small and isolated patches of cultivation scattered over it.'[7] A 'village' represented a broad area in which two or three families might reside for a few years before moving on. A record of such a 'village' called Khapadya exists for the period 1830–70, which gives an idea of the way in which a locality was occupied and later deserted. In 1830 the area was being cultivated by a Konkana called Jana Patel. After the Dangs was conquered by the British in 1830, he deserted the place. It remained uninhabited for many years. Eventually Jana returned with his son Eru, now aged 35. However, as they were forced to provide labour for touring officials and others from outside the Dangs, they soon left and went elsewhere. In 1867, Janu, Eru, and three other Konkanas returned to settle the site once more. In 1868 cholera broke out and Janu and some others died. The rest, including Eru, deserted the village. It was still uninhabited in 1871.[8]

Despite the fact that they had to pay taxes to the chiefs and satisfy other demands from the Bhils, the Konkanas were normally able to eke out a living without having to rely on hunting and gathering. The kumri cultivation of the Bhils and Varlis, on the other hand, provided grain for only two or three months after harvest; during the rest of the year they depended on other sources of subsistence. They ate forest fruits, *mahua* flowers, grain from wild grasses, and various roots. The Bhils were also keen hunters with the bow and arrow and, in a few cases, guns. They would set fire to the tall grass to drive the animals out so that they could sight them and kill them. Hares were smoked out of their holes or trapped in nets. They also caught and ate fish and birds.

The Forest Economy of the Dangs

The Dangs was well-integrated into the economic life of the outside world. Forest produce was in demand, and the people of the tract could earn both money and payment in kind by providing it. They did not cut and sell timber, for they were not able — or indeed prepared — to transport huge tree trunks up the Ghats to Khandesh or down to the richer markets of south Gujarat. Large teams of bullocks to drag the timber through the forests, and sturdy bullock carts, often drawn by four bullocks, were required to transport the wood across the plains to the market. This trade, which thrived before the coming of the British, was organized by merchants who lived in the towns surrounding the Dangs. The normal pattern was for a trader to contact a Bhil chief in whose area he wished to operate. Once permission had been obtained, peasants with bullock teams and carts were hired from the plains. These peasants were free from their own agricultural work between February and May, and during this period they went to the Dangs. The wood was cut by them, and the local chief was paid a sum for each cartload exported. The payment was made both in cash and in kind (e.g., grain, cloth, and other products not obtainable in the Dangs). The levy on timber provided a major source of income for the chiefs — greater than other sources, such as the tax on subordinate peasantry or the protection money they levied on the surrounding plains. For instance, in 1825, Silpat had an annual income of about Rs 5000. Of this, less than Rs 1000 came from protection money,[9] and still less from land tax.

The Dangis sold and bartered other forest produce to merchants who lived on the borders of the Dangs or to the Banjara pack-bullock carriers who came to the forest to graze their cattle during the summer. In 1843 the trade in mahua flowers was reported to be second only to that of teak from the Dangs.[10] On 10 April 1856, a forest official travelling in the Dangs reported that the flowers of mahua were seen in full bloom and that the entire population was out gathering them for sale.[11] Some of the crop was carried by the Banjaras to the plains; the rest was taken to the borders of the Dangs and sold to liquor dealers who used it for making country liquor to be marketed in the villages and towns. For many Dangis, the mahua trade provided the chief means for earning cash, which they used for buying clothes and other essentials. The Dangis also made bamboo products — such as baskets, bamboo matting, and winnowing mats — which they then took to the borders of the Dangs. Traders purchased these products and sold them in the plains. Other forest products sold in this manner were honey and lac.

Dangis kept livestock, though they were not important for providing food directly. The Konkanas used bullocks mainly for ploughing and cows for breeding. The cattle were of small stature, and the cows did not provide adequate milk for the people. Very few buffaloes or goats were kept. The cattle were let out to graze in the forest. In addition, Banjaras came into the Dangs during the hot weather to graze their herds, which could consist of up to 500 animals. They paid the Bhil chiefs for the right to graze — usually a charge fixed per 100 cattle. They often brought with them a little coarse cloth, cheap jewelery, beads, earthen pots, and salt, which they sold or bartered to the Dangis.

The relationship between the Dangis and the outside world had its more violent side. The Bhils levied protection-money from the rulers of the surrounding plains, on threat of looting if the money was not paid. In the early 1820s, for instance, they began raiding villages in the territory of the Gaikwad of Baroda lying to the north of the Dangs, claiming that the Gaikwad's officials were refusing to hand over their customary tribute. Silpat claimed that protection-money worth Rs 1520 a year was due to the Bhil chiefs, whereas the Gaikwad was prepared to concede only Rs 940.[12] This dispute led the Gaikwad to send some troops into the Dangs in 1825 to crush the Bhils. The Gaikwad suffered a humiliating defeat, being driven from the Dangs with casualties.

This evidence brings out the complexities of the forest economy of the pre-colonial period in the Dangs. It is difficult to argue that

forest dwellers such as the Dangis were in any way 'natural conservators' of their environment. For the people of the Dangs the
forest was eternal; there was no consciousness that human beings
could destroy this massive expanse of vegetation. They used the
forest to maintain their way of life, and the Bhil chiefs profited
by allowing outsiders to exploit it on their own terms. Clearly,
however, they had a strong affinity with these woods and hills —
their home as well as place of refuge — and any destruction that
they carried out was small-scale and made very little difference to
the environment as a whole. With the coming of the British, a
great change was initiated in which the forests — and the people
who lived there — were to become managed as a resource within
a wider system of production.

The Erosion of Tribal Forest Control
in the Early Colonial Period

The British conquered the Peshwas and occupied the region to
the east of the Dangs in 1818. The district of Khandesh was
established, with its collector being made responsible for the
Dangs. However, it was only in 1830 that British troops were sent
into the Dangs to subjugate the Bhil chiefs who had from time
to time raided villages in Khandesh. The chiefs were recognized
as independent rulers under overall British paramountcy, with
their protection monies being paid regularly by the colonial state.
No attempts were then made to rule the Dangs in any direct way.
Many British soldiers had died of malaria after the 1830 expedi-
tion, a mortality that revealed the dangers of living in those hills,
even in the 'healthy' season of the year from March to May.
Instead, the British relied on a system of informal control, using
officials known as the 'Bhil Agents'. The chief role of the Bhil
Agent was to keep the Bhils quiet and out of trouble.

The Bhil Agents sought to encourage the Bhils to settle down
and learn to subsist from settled agriculture rather than from
raiding, hunting, gathering, and shifting cultivation. In the words
of one Bhil Agent: 'I have always made it a rule when visiting Bhil
settlements to personally go and see for myself how the Bhil is
getting on with his cultivation, and have endeavoured with kind-
ness and quiet persuasion to encourage him in the steady and

peaceful occupation of agriculture.'[13] As early as 1834 the Khandesh authorities tried to encourage merchants to enter the Dangs and provide loans to the Bhils to enable them to carry on settled agriculture. They also sought to encourage peasants to migrate into the tract from outside to settle and cultivate the land. The emphasis here was on clearing rather than on preserving the forest.

Left to themselves, the Bhil Agents and authorities in Khandesh would have been happy to continue this loose relationship with the Dangi Bhils, not forcing the pace of change in the tract in any dramatic manner. The thrust for a more active intervention within the Dangs came from the British navy which needed high-quality teak for the construction of warships. During the early years of the nineteenth century, the navy had obtained such wood from the forests of the Malabar coast, but overcutting led to dwindling supplies in the 1830s. Increasingly, they looked to the forests of south Gujarat. Initially they purchased wood from local timber merchants who had made agreements with the chiefs to cut the wood. It was felt, however, that prices were too high, and in 1843 the Bombay Government decided to start extracting timber directly. The chiefs were forced to sign leases handing over their rights to all timber from the Dangs. They were compensated with a relatively small annual payment.

The chiefs did not initially understand what they had signed. There is a strong oral tradition in the Dangs that the British tricked the chiefs into taking money, not knowing that this entailed the loss of their rights to use the forests. When the full significance became clear, they tried to repudiate the leases and returned the first payments. In the words of Anandrao, chief of Vasurna: 'No interruption [to our control of the forest] was ever experienced by us during the Mugal and Peshwa's Government, nor under the British till now.'[14] The chiefs forced the first woodcutters, whom the British sent into the Dangs, to stop their work. To forestall further trouble, the Bhil Agent agreed to raise the annual payments slightly but insisted that the British be given full and sole power to cut, conserve, and plant timber in the Dangs. The chiefs were allowed to cut timber for their own use only. The government would establish customs posts and levy taxes on timber and other articles that might be taken from the tract. The government was allowed to clear land in the Dangs and provide for its cultivation, though the tax on cultivation was to go to the chiefs. The government was permitted to improve roads to facilitate the export of timber. The leases could be relinquished by the government at a six-month notice, but not by the chiefs. After sixteen years the government would be allowed to take out new leases on terms they might judge as equitable. The chiefs were required to give full assistance to the government timber agents. If local people were asked to cut timber, they were to be paid the prevailing rate by the timber agents.

The leases that the different chiefs were forced to sign were one-sided. The sums granted in compensation were meagre when compared to the value of the wood. As an official later gloated: ' . . . such leases [were] . . . obtained at a marvellously cheap rate from the rude and uncivilized owners by means of persuasion.'[15] The new leases allowed the chiefs to take wood from the forests for their personal use; there was no provision for the rights of the inhabitants of the Dangs as a whole. The chiefs were allowed to collect land tax, but this was because the amount was small and the British would have found it troublesome to collect it themselves. Although the leases were limited to sixteen years, the limit had no significance as the British insisted on having the sole right to renew the leases on whatever terms they pleased. The Bhils had no power to revoke the leases at any future date; this privilege lay entirely with the government.

From this time onwards, the power of the forest officials gradually began to eclipse that of the Bhil Agents in the Dangs. In contrast to the latter, the forest officials placed much emphasis on the utilitarian virtues of hard work and careful husbandry. They viewed the forest as a resource with two dimensions to it. First, the forest officials considered it as a property they were holding in trust for the chiefs and from which they were entitled to take an income. Their periodic statements of account showed the cost of administrating the forest (including the amounts of lease money paid to the chiefs), which were set against the amount derived from the sale of wood and the sale of permits to cut wood. When this balance showed a loss (which was rare), noises were made that the Dangs were 'not worth keeping'.[16] On the whole, however, the 'capital' invested by the British earned large dividends. It was the duty of forest officials to 'protect' this resource from the ravages of the indigenous inhabitants. According to one official, the forests of the Dangs were 'a sort of national

1. Eucalyptus saplings, while fast growing, generate few non-timber forest products, do little to enrich soils, or rebalance hydrological systems (M. Poffenberger).

2. Chingra village leader and local forest guard kneel proudly in front of their regenerating sal (*shorea robusta*) trees. The biologically rich understorey comprises over 200 species, more than 70 per cent of which are utilized by villagers (M. Poffenberger).

3. Small tribal shrines dedicated to the smallpox goddess offer animal figures, commonly found in sal forests. Spirit forests protect mother trees and seed sources, creating small biodiversity reserves assisting the regeneration of degraded forests (M. Poffenberger).

4. Chingra villagers pose for a photograph after discussing forest protection activites (M. Poffenberger).

5. Santhal tribal boy with a bow and arrow used for hunting and carried on forest-protection patrols (M. Poffenberger).

6. The removal of sal and other tree root stock, combined with heavy grazing pressures cause topsoil loss and compaction. This suppresses and undermines possible forest regrowth (M. Poffenberger).

7. This scrubby vegetation is actually mature sal trees reduced to scrub from constant fuelwood hacking. Under effective community protection, within five to ten years it will be transformed into a dense, closed canopy, multi-tiered sal forest (M. Poffenberger).

8. The resistance of the Santhals to forest alienation is depicted in this lithograph from the *Illustrated London News*, 23 February 1856.

wealth, and it behooves the state to devise means whereby its future supply may not diminish by injurious treatment of the trees at the hands of an improvident class of people.'[17] Another official likened the forest to a form of capital from which the society drew interest and which was being threatened by its inhabitants:

It is the destructive attacks made upon the capital, dissipating the latter, and rendering the continuance of interest impossible, which the Forest Officer, from his professional education, perceives when it is not apparent to the uninitiated, and which he finds so sore a burden on his conscience.[18]

Such a view was wholly alien to the people of the Dangs; for them the forests were a way of life and a spiritual home, not mere timber with a capital value. Second, the officials viewed the forest as a 'natural resource' that the government had a duty to protect. Naturalists had for long realized that forests occupied a key place in the natural cycle. Since classical times, writers had observed that whenever there was a loss of forest cover, rivers dried up during droughts and flooded badly when there was heavy rain, carrying away topsoil. This knowledge was given a scientific base in the late eighteenth and early nineteenth centuries. Forests were seen to encourage rainfall, check run-off, prevent floods, and help water to percolate into the ground and be available from perennial springs. During droughts, they provided succour to the people who could come to the forest to search out wild roots and fruits, while hungry livestock could be brought to eat dry grass and leaves. The voice of conservation was always weak as opposed to the much stronger imperial need for wood and arable land which could grow food grain and provide a higher income to the state from land tax. The official argument was, however, important and valid, and the government paid attention to it by creating forest reserves during the second half of the nineteenth century.

The colonial officials' view of conservation was, however, gravely flawed. They believed strongly that forests could be saved only if managed in an authoritarian manner by foresters. They also believed that left to itself, nature degenerated. According to Donald Worster: 'That impulse became a defining theme among Victorians, much as it had been for their reforming Protestant forebears. While they went on loving their gardens and city parks, where *they* were in control, they were determined to dispel any foolish notions about the innate goodness of natural forces at work on earth.'[19] It was the duty of the colonial official to tame and control the natural world.

Human beings who lived close to nature were also seen as a threat. It was believed that civilization could advance only with constant application, effort, and exertion. Otherwise, humans remained in a slough of savagery, using resources wastefully, and causing greater degradation to their environment. For the Victorians such savagery was thus 'an offence that ought to be crushed out wherever it was found.'[20] Forest officers believed that they had to police the forest to protect it from the 'primitives' who lived there. Their relationship with the forest dwellers was thus one of antagonism rather than co-operation. This attitude prevented them from learning how to work with them to protect the forests. If they had done this, their task of conservation could have been carried on with much less friction and with a far greater chance of success.

A further problem with forest conservation, as it developed into a profession in Europe in the nineteenth century, was that foresters were connected closely to hunting. According to Jack Westoby, 'There has always been an association between landed property, forest estates, foresters and hunting . . . At one time many young men chose the forest profession precisely for this reason.'[21] Foresters saw their role as not only protecting timber but also as maintaining hunting grounds for the recreation of the rich. Most foresters in India were also keen *shikaris*. This also brought them into conflict with the people. For example, hunting of tigers caused much distress in the Dangs, as the animal was revered there as *Vagh Dev* (forest spirit/god). The Dangis did their best to avoid being involved in such slaughter. The forest dwellers also disliked performing the dangerous task of beating the drums to drive wild animals toward the guns of the hunters. In some cases, tribals fled their homelands when the local *rajas* forced the villagers to perform these tasks.

In the Dangs these contradictions in attitudes came into sharp conflict only towards the end of the nineteenth century, when the forest officials took active steps to 'save' the forests from the people who lived there. For the first fifty years of the forest leases, control over the day-to-day life of the Dangis was slack, for no officials lived permanently in the Dangs and there were hardly any roads to allow easy movement into the tract. There was, however, a consistent and rising exploitation of the forest.

During this period the demand for teak for shipbuilding was superseded by a much larger demand for teak for railway sleepers. Initially this new demand was met from the thinner and more accessible forests of Thane and Surat districts, and the princely states of Dharampur and Vansda. By the late 1850s these forests were showing signs of exhaustion, and timber dealers were demanding that they be given access to the Dangs. Since the signing of the leases, the British had taken direct control of cutting, hauling, and selling timber from the tract, using peasant labour hired from the plains. During the 1860s they tried out a system of giving a timber contract to supply sleepers for the Great Indian Peninsula Railway from the Dangs to a private dealer. The contractor invested his own capital in building roads to carry wood from the tract. But, in 1869 the British changed to a system in which the right to cut and transport timber from a stated forest-area was auctioned to timber contractors for a particular period. The variety and the amount of wood to be cut was mentioned in the contract. The idea was that the forest officials would single out trees ready for cutting and mark them. The contractor then hired cutters to go into the forest to fell the trees. The cut timber was then taken to checkposts on the borders or to the depots outside the Dangs, where it was measured, its value determined, a stamp put on it, after which it could be removed. The contractor had to pay the government the sum agreed for each variety of wood thus extracted. To help the contractors, the British constructed a new road into the heart of the Dangs, which provided access to fresh tracts of uncut timber. As a result, the volume of timber exported rose considerably during the 1870s.

Thereafter, forest officials became more and more strident in their demand for better 'protection' of the forest from the people who lived there. Without much logic, the onus for its destruction was placed on the Dangis rather than on the system of timber extraction operated by the government. One practice considered particularly harmful was the annual burning of the forest floor to clear the tall grass that harboured wild animals and made hunting and travel through the forest difficult. Large numbers of saplings were supposedly destroyed in these fires so that regeneration of the forest was prevented; existing trees were damaged, so that they grew crooked and became 'valueless' as timber. Slash-and-burn cultivation was also blamed. The lopping of branches for rab and kumri cultivation was considered to damage trees. The Bhils were accused of cutting and then burning large tracts of choice forest. The scanty crop that they raised was not, according to forest officials, worth the value of the smallest tree cut.

Consolidating Government Forest Controls in the Dangs

From the 1870s, officials of the Bombay Forest Department began to demand that the forests of the Dangs be managed along the same lines as prevailed in the Bombay Presidency. The Forest Act of 1865 created tracts of 'reserved' forest in India in which the government could ban all cultivation. The government held full rights of ownership over these reserves.[22] The rules were tightened by the Forest Act of 1878, which held that peasants who lived in the reserved areas had no 'right' to the forests. At best their use of it was a 'privilege' granted by the state.[23] During this period the forests of Bombay Presidency were surveyed, the 'privileges' of the existing inhabitants listed — in a way depriving them of most of their customary rights — and boundary posts were erected to demarcate the extent of the area reserved under the terms of the Forest Acts. Forests that had not yet been surveyed, but which, it was hoped, would be reserved in the future, were classified as 'protected.' The forest officials in charge of the Dangs demanded that the tract be similarly surveyed and divided into reserved and protected zones. The difficulty was that the Dangs were legally under the Bhil chiefs, and the laws of the Bombay government did not apply there. They could not therefore be brought automatically under the operation of the two Forest Acts.

In 1878, the Bombay government decided to look into the problem by appointing a commission of inquiry consisting of political as well as forest officials. The final report of the forest officer stated:

For years past, the wasteful system of cultivation carried on by the inhabitants of the Dangs, and their consequent migration from one place to another, on the appearance of cattle disease, or of a reputed witch, or in consequence of the death of a member of their community,

or for any other absurd reason, have undoubtedly tended to the destruction of the forests. We are, therefore, of opinion that efforts should now
be made, at any rate, to restrict such destruction, in order to ensure a
fair supply of valuable timber for the future. This can only be done by
demarcating forest tracts and, as far as possible, separating these from
the area actually required by the people for cultivation.[24]

It was recommended that agriculture be concentrated only on
land suitable for perennial cultivation, with a constant supply of
water nearby. Settled colonies needed to be established on this
land. The rest of the Dangs would then be designated as reserved
forest in which grazing would be allowed, but no cultivation. The
recommendations were accepted by the Bombay government in
1879. It was ruled that about half of the land area of the Dangs
should be demarcated as reserved forest in which cultivation would
not be permitted.[25]

Although the principle of forest reservation for the Dangs was
thus accepted, more than a decade passed before reserved areas
were actually demarcated. This was partly due to technical difficulties; the Dangs had never been surveyed thoroughly; more
importantly, the Bhil chiefs opposed the entire idea vociferously.
They argued that the British had no right under the terms of the
leases to limit cultivation to particular areas of the Dangs. The
authority of the chiefs over other Dangis was connected closely
with their right to determine who cultivated any particular part
of the area under their rule. They also valued their own right to
move about in, live in, and use their forests as they pleased. The
idea of being confined to particular areas angered the Bhil chiefs.
For many years they had faced growing harassment from lower-
grade forest officials, and they knew that once the forests were
reserved this problem would become worse.

The British countered by promising to increase payments to
the chiefs under the terms of the leases. After much forceful
persuasion by the British, the chiefs agreed to the principle of
reservation and signed the following agreement in 1889:

I understand that Government wish to reserve for forests a portion of
my Dang, not exceeding one half, and that if this is done in the portion
reserved, no one would be allowed to enter without permission from
the forest officer. I am willing to agree to this arrangement on the
understanding that sufficient land for cultivation will be excluded from

reserved forest. I and my people will be allowed to use the trees outside reserved forests for agricultural and domestic purposes, but not for sale or barter.[26]

The demarcation surveys were carried out in 1891 and 1892. The chiefs did their best to keep as much land as possible outside the reserved area. In the end, after many compromises, 34 per cent of the land area was designated as reserved forest.[27]

The chiefs resented the loss of even this amount of their forest, and actively encouraged the peasants of the Dangs to cultivate in the reserved rather than in the unreserved areas. They also connived at the wanton destruction of the forest. Damaging fires were started and many trees were cut and destroyed. In 1901 one official reported that he saw large numbers of stumps of recently cut trees. Forest, which eight or nine years before had been covered with thick, fine teak and bamboo, was now scarcely worth the trouble of conservation; 'the deterioration has been going on at a terrible pace since the demarcation was made and particularly since 1896.'[28] Forest officers who tried to prevent this destruction were threatened and even beaten by the Dangis. During the period of famine and death between 1899 and 1902, this discontent and the starvation distress triggered off a wave of looting of the subordinate peasantry by the Dangi Bhils. Particularly notorious was a gang of some relatives of the Raja of Amla who carried on their raids from a base on the Atalya hill. A force of more than 200 men was sent eventually to suppress them, and only after most of them had been captured or driven underground did peace return to the Dangs in 1903.[29]

On 1 January 1903, responsibility for the administration of the Dangs was transferred from the authorities in Khandesh to the forest department of the Bombay Presidency. Thereafter, the Bhil Agents ceased to have any power in the tract. The divisional forest officer for Surat Circle, in which the Dangs were the chief forests, became the assistant political agent responsible for the day-to-day administration of the Dangs. He was accountable to the political agent, who was also the Collector of Surat. The assistant political agent was given judicial powers to try all but the most serious criminal cases in the Dangs. E.M. Hodgson was appointed to this post, serving in it from 1903 to 1910.

Hodgson started by constructing the headquarters at Ahwa, a

village on a plateau in the heart of the tract. Besides building a forest rest house for white officials, houses were built for other officers, police officers, and servants. A small dispensary, a jail, a post office, and a storeroom were constructed. Hodgson also reorganized the administration. He drew up a code of administrative standing orders, which conformed as far as possible to the laws in force in the Bombay Presidency, modified to the particular circumstances of the Dangs. He organized a police force with fourteen constables based at Ahwa. Roads were improved, allowing better communication with Ahwa. Forest rest houses were constructed at various strategic points, and more wells were dug to provide clean drinking water.

Hodgson reclassified certain tracts in which cultivation could not be prevented as 'protected forest'. There were therefore three categories of land in the Dangs: reserved, protected, and open. He banned all cutting of live trees in the reserved forests and of immature trees in the protected forests for ten years, to allow the forest to regenerate after the devastation of the previous decade.[30] Due to the greater efficiency of the administration and the opening up of the Dangs through better communication, the revenue from the forest did not suffer from these restrictions. In fact, it increased.

Hodgson also tried to stop the annual burning of the forest floor. In the 1890s, officials had despaired of ever being able to prevent these fires. Hodgson began by trying to persuade the chiefs to dissuade their followers and subjects from starting fires. He promised to give rewards and turbans to minor chiefs and village headpeople who managed to prevent fires. He burned strips of forests to create firebreaks to prevent fires from spreading all over the Dangs. Vigorous efforts were made to catch and punish anyone causing fires. In 1907 a complete ban was imposed on the burning of grass under mahua trees, the smoking out of rats and other animals, and the use of fire torches to provide illumination at night. Liberal payments of rewards and distribution of turbans were made to those who co-operated. As a result, whereas in 1905–6, 93 per cent of the Dangs had been burned, one year later only 17 per cent had been burned.[31] This proved to be the turning point; thereafter, the incidence of fires remained low.

The seeming success of this policy masked a continuing resentment among the Bhils against Hodgson's policy. They continued to regard the entire forest as their preserve, to be used as they pleased. They disliked settled agriculture and resisted being made to live in the valleys and cultivate fixed plots of land. Many continued to cultivate where they wished, which led to inevitable conflicts with the forest guards and police. Deprived of their livelihood, some of them took to extracting food and other goods from the subordinate peasantry on a more regular and exploitative basis. The subordinate peasantry then looked to the police for protection, which led to further conflicts. These tensions gave rise to three uprisings by the Bhils between 1907 and 1914.

In the first of these, in September 1907, some Bhils attacked Ahwa, smashing all government property they could lay their hands on. Files and books were torn and furniture was broken to pieces. Trees in the compound were uprooted. They forced their way into the dispensary and broke the surgical and medical instruments and smashed the bottles of medicine. In 1911, some Bhils of Kadmal village protested against the forest department by setting fire to large tracts of forest. Forest guards sent to Kadmal to inspect the damage were threatened by the Bhils; to save themselves, they handed over a large number of licenses to cut wood free of cost. The defiance spread, and the forest was burned elsewhere. In Ahwa there was a mood of alarm, and a feeling that the events of 1907 were about to be repeated. Police reinforcements were sent from outside, and the ringleaders arrested. In this outbreak, the leaders against the forest regulations were not the top chiefs but some of the lesser chiefs. They received firm support from the majority of the Bhils. Their object was to reassert the authority they had lost — demanding that the forest guards stop interfering with their use of the forests. The Bhils did not receive support from the subordinate peasantry. The latter helped the authorities when they could and lay low when the Bhils went on a rampage. The Bhils clearly resented that many Konkana village headmen were assisting the British, and they were also often singled out for attack.[32]

A further outbreak occurred in December 1914. This was set off by rumours that the British were facing imminent defeat in the European war. The Bhils saw this as an opportunity to reassert their authority. Large bands of Bhils armed with bows and arrows assembled under the leadership of the Naik of Pimpri and rajas of Gadhvi and Amla. They set about burning the forest and cutting down trees. Forest officials who tried to stop them were threatened with violence, though none was actually harmed. The British took quick action, sending a large body of armed police to the Dangs, who soon restored order. Those who had taken part were fined or imprisoned. Four chiefs who had remained aloof and many Konkana Patels who had helped to restore order were given rewards of bangles and money.

Increasingly, the peasants were being made to settle down to cultivate the land demarcated for agriculture. They were encouraged to stay on fixed sites, construct permanent fields, and even make terraces for paddy cultivation. They were helped — through government loans — to buy carts and bullocks to make up a plough team. They were ordered to use a mixture of cow dung, grass and forest debris for their rab, rather than tree cuttings, as before. A few small dams were constructed by the government to store water throughout the year and to carry out irrigation. The peasants were told that they should grow 'better' (i.e., more marketable) food grains such as *juvar* (sorghum) or rice, and to keep stocks of grain as well as fodder. Fruit trees, such as mango and tamarind, were planted near villages to encourage the peasants to stay in one place. The money spent on these activities was not, however, great, and the impact was limited. Although the peasants were forced to keep out of the reserved forests, they refused to radically change their methods of farming.

The final years of the colonial rule were ones of great hardship for the people of the Dangs. The forest department ruled the tract with a heavy hand, forcing the people to carry out paid labour for the forest department and much unpaid labour for the touring officials. Although the paid labour provided some source of income to replace that lost from selling forest produce, the Dangis continued to resent having to do such work. Forest guards were free in their use of punishments to keep the people from using the forest reserves.

During World War I there was an urgent demand for timber, which led to its large-scale cutting and extraction. In 1914–15 the timber income from the Dangs was Rs 154,133; by 1918–19 it was Rs 460,021. The chiefs themselves continued to be paid their lease money. In 1924 the payments were raised by 50 per cent to a total of Rs 26,921. This increase however, did not adequately reflect the amounts being earned from the Dangs: during the 1920s the average annual profit to the British was, Rs 154,917 per year. By the 1930s the chiefs were demanding that their payments be increased, considering the profits being made. The Bombay government turned down this request, arguing that if they received more money they would merely squander it on drink.

After the failure of the 1914 revolt, the Bhils did not dare to risk any further confrontations with the colonial state, but resisted only surreptitiously. They maintained their self-respect in a period of adversity by taking pride in their claims of royal lineage and by keeping alive the memory of the times when they had once ruled the forests.[33] They tried to deny the legitimacy of the loss of their forests by repeating stories of the devious deceptions by which — it was said — the British had swindled them of their inheritance. Likewise, they continued to celebrate their revolts against the British during the early years of the twentieth century. In 1949, encouraged by the news that the British had left India, they staged a further revolt, burning large areas of forest. The revolt, however, was suppressed quickly by the police of independent India.[34]

Forest Management after Independence

When India won independence in 1947, the Dangs District was opened to political activists and social reformers who had until then been excluded from the tract. The ground had already been prepared for them by a few Konkanas who had started Gandhian-style activities — such as spinning, weaving, anti-liquor and anti-meat activities — on their own initiative in the years preceding independence. In 1948 the brothers Chhotubhai and Ghelubhai Nayak came to the Dangs and established a Gandhian *ashram* at Ahwa. A large numbers of Dangis flocked to the ashram with complaints against the forest department, many of which related to sexual harassment by the forest guards against adivasi women. The two Gandhians accepted these grievances but advised the people that in the long run, they could only improve their position through education and social reform. They started primary schools, taught people to spin and weave cotton, and persuaded many to give up drinking liquor. In 1949 they started a multi-purpose co-operative that, among other things, distributed grain to members. The Bombay government agreed to give this co-operative the right to cut wood in certain forest coupes, the aim being to undercut some of the power of the forest department in favour of the people.[35] This was the start of a development that was to have a profound impact on the Dangs.

The first forest labour co-operatives in Bombay State were started in Thane District in 1947 as part of a strategy to counter the strong communist agitation there against landlords, money-lenders, and forest contractors. The Gandhian, Jugatram Dave, started similar forest labour co-operatives in Surat District in 1948, the intention being to break the hold of powerful forest contractors who had always been allied closely with the British. The co-operatives proved a success: those who cut wood through them soon found their work conditions much improved. No longer were they exploited and humiliated by powerful contrac-tors, and their earnings were also higher. The co-operatives pro-vided a strong political base for the Gandhian activists who headed them, and they used the co-operatives to propagate their ideas.

In 1950, six new forest labour co-operatives were established in the Dangs. They soon, however, became embroiled in a political controversy revolving around the issue of whether the Dangs should form a part of Gujarat or Maharashtra. Being a border region, with local dialects consisting of Gujarati as well as Marathi elements, and a tradition of rule by both Maharashtra (Khandesh) and Gujarat (Surat), the status of the tract was ambiguous. During the twentieth century, the ruling authority had been based in Surat, and the Gandhian activists who became the first new-style political leaders in the area were Gujaratis. Education in their schools was in Gujarati. Maharashtrians were not, however, pre-pared to accept this. Two Gandhians of Nasik founded a new institution, the Dang Seva Mandal, which opened schools in which Marathi was the medium of instruction. They persuaded the Bombay government to give some forest coupes to the forest labour co-operatives they had started. However, in 1954 the Bombay government discovered certain financial irregularities of the Gujarati societies, which led to their liquidation.

In 1956, as a result of strong agitation in Maharashtra, the Government of India announced that Bombay state would be divided into two separate linguistic states. The actual division did not, however, take place for another four years, and the status of the Dangs continued to be uncertain. In the Dangs, the two rival Gandhian groups fought for the Dangs to be included in either one or the other state. By this time there were twenty forest co-operatives working in the Dangs, all loyal to the Gandhians from Maharashtra. They made large profits, and much money was reinvested into organizing the agitation for demanding that the Dangs go to Maharashtra. There were frequent meetings and demonstrations. Leading nationalist politicians of Gujarat came to the Dangs to counter this propaganda. Chhotubhai and Ghelubhai Nayak founded 179 primary schools to propagate Gujarati education. In 1958 the first election was held for the new Dangs district *panchayat*. The election was contested fiercely by the two groups. The Gujaratis won twenty-six out of thirty seats, and Chhotubhai Nayak became the president of the district panchayat. It was largely because of this result that when Gujarat and Maharashtra were divided in 1960, the Dangs were awarded to Gujarat.

The Maharashtrian forest labour co-operatives were abolished in 1962 by the new government of Gujarat, and until 1967 all cutting was carried out by either the forest department or contractors. The co-operatives were revived in 1967. However, as soon as Chhotubhai Nayak won the award of forest coupes for the two societies he had started, his political rivals from other parts of south Gujarat countered him by establishing other societies under their control. As his rivals did not reside in the Dangs, they encouraged some young Konkanas to take over the day-to-day management of their co-operatives. The leading figure among this rival group was Ramu Thakare, a Konkana of Sarva village who had served for a time in the Indian army before returning home to the Dangs in 1967. He consolidated his position rapidly, and in 1970, using the co-operatives as a political base, managed to oust Chhotubai Nayak from the presidentship of the district panchayat. He has remained president to this day. The young Konkanas quickly captured control over Chhotubhai's co-operatives and founded some new ones, bringing the total to thirty-two.

Between 1967 and 1986, forest cutting throughout the Dangs was controlled by these highly politicized labour co-operatives. They built a network of power, which revolved around forest employment. With the continuing growth in population, it became increasingly hard for most Dangis to make an adequate living from cultivation. Between 1931 and 1991, the population of the Dangs grew from 33,750 to 1,43,500. The only other major source of livelihood was forest work. The power to award such work lay with the leaders of the co-operatives and their agents in the villages. This power allowed them to build vote-banks and win seats for themselves on the district panchayat. This, in turn, gave them control over the allocation of government development funds — power that was often used corruptly. Many of these managers and agents of the co-operatives were suddenly seen to acquire new farmlands, large houses, and personal jeeps. Ordinary adivasis found that they had to bribe these new leaders to obtain forest work or to have any government policy implemented in their villages (e.g., to obtain government loans for land improvement). Very little of the government money earmarked for developing the Dangs actually reached the poor, although income from the forests of the Dangs increased from Rs 12,50,000 in 1970 to Rs 98,60,000 in 1985.[36]

The increase in forest income reflected to some extent an increase in the exploitation of the forests. Both the Gujarat government and the leaders of the co-operatives were interested in maximizing earnings from the forest. The latter had little interest in conservation; their main concern was to earn as much as possible while they held power. The rise in income was compounded by the soaring price of wood. But just as important, it reflected an increase in the area under reserved and protected forest. For many years the forest department had been gradually encroaching on areas in which cultivation was permitted, which included a good deal of protected forest, and reclassifying them as reserved forest. They were able to do this because most Dangis were still cultivating on a shifting basis in an area around their, by now, fixed village sites. Land, which was not being cultivated in a particular year, could be planted with saplings by forest employees and reclassified as forest.

In the late nineteenth century, about 66 per cent of the Dangs was classed as available for cultivation. By 1956–57 the figure had fallen to about 54 per cent. During the 1960s, the Gujarat government conducted a survey of the Dangs, aiming to give land-occupancy deeds to the cultivators. As a result of this survey, 79,913 hectares were classified as farmland, which represented 45 per cent of the total area.[37] Much of the cultivated land was omitted from this figure, as it was not being farmed at the time of the survey. Land deeds were distributed in 1970, on condition that the forest department would continue to own trees on the land and that the occupants could only lop branches for rab and not fell trees. No forest produce could be taken from the reserved forests, which by now occupied about 50 per cent of the Dangs. During the next decade, the forest department continued to encroach on farmland, often by planting trees against the wishes of the occupants. In 1981 only 51,186 hectares were recorded as cultivated land, or 36 per cent of the Dangs.[38] It was claimed that the land had been surrendered voluntarily, which was hardly credible in view of the rapid growth in population, creating a severe land hunger throughout the tract. These developments benefited chiefly the Gujarat government and the leaders of the labour co-operatives. Both had a vested interest in increasing their income by extending the forest and enforcing forest laws. As the farmlands of the Dangis were interspersed with tracts of reserved forest, it was all too easy for a peasant to break these laws by, say, allowing cattle to wander into the reserve. Anyone suspected of taking wood or other produce from the reserves was arrested and fined. Often, the only way of escaping such punishment was to bribe the forest guards who were nicknamed *kam-da chor* (chicken thief) for habitually demanding free chickens to look the other way.[39]

Contemporary Community Forest Control Movements

From the late 1970s, the people of the Dangs began to assert themselves against the forest officials and corrupt politicians. The chief demands — as recorded in a 1977 meeting — were that they be given a right to fell any tree on their land, that they be allowed to take bamboo from the forests, that land taken by the forest department for tree plantation be returned, and that a new survey of the Dangs be conducted to correct the shortcomings of the 1960s survey. A new organization, the Gram Vikas Mandali, was started to voice these demands. The founder of this body was an Australian social worker, Barry Underwood, who was based in Ramu Thakare's village of Sarva. He enjoyed the support of some educated adivasi youths who felt that the forest officials and the leaders of the forest co-operatives were stifling the social and economic development of the Dangs. One of their chief demands was for more employment for educated adivasis. Barry Underwood, in contrast to the Gandhians, ran the organization democratically, encouraging the youths to be directly involved in decision making. Because of this, the organization not only continued to flourish after he left the Dangs in 1983, but extended its activities to adivasi women.

In 1986, alarmed at the rapid destruction of forests throughout India, the government imposed a moratorium on all cutting of trees in forest reserves. Thus, the forest labour co-operatives in the Dangs lost their chief reason for existence. The only work left involved conservation activities, and there was comparatively little employment in this sphere. Large numbers of Dangis were forced to migrate outside the Dangs to obtain seasonal agricultural employment in the plains of south Gujarat. Within four years of the moratorium, at least one-third of the entire population left the tract as soon as the monsoon was over, returning only in summer, before the next monsoon. Working conditions in the plains were extremely bad; the pay was very poor, and although the migrants could obtain a bare subsistence by such means, they could send back home very little money and often returned at the end of the season almost empty-handed. On the whole, the younger and fitter people migrated, leaving the elderly behind in a state of almost complete destitution.

Active resistance by the people of the Dangs began in 1989, when a dozen forest guards went to Gira-Dabdar village to confiscate wood that the villagers had allegedly taken from the reserves. Village women surrounded the guards and told them forcefully that the forest department had no right to the wood. They argued that the forests had been taken from the Dangi people by the government through trickery and that they had a moral right to earn their livelihood by using forest produce. The guards were forced to retreat. These women had been involved in the activities of the Gram Vikas Mandali, for which Gira-Dabdar was a stronghold. The event led to more a organized and planned resistance. The leaders who now emerged were two outsiders — Irfan Engineer, a Naxalite from Bombay and Virsinhbhai Patel, a Chodhri adivasi from Songadh, in Surat District. In November 1989 they started an organization called the Adivasi Bhumihin Kisan Hak Rakshak Samiti (Association for the Protection of the Cultivation Rights of Adivasis). This built a strong base in twenty-one villages in the western Dangs where the Gram Vikas Mandali had already been most active. There were members in an additional forty or so villages. Their chief demands were that a new survey of agricultural land in the Dangs be conducted and that land be made available in the forest reserves for agriculture.

The Bhils had, since the creation of the forest reserves in the late nineteenth century, refused to accept that the state had a right to prevent cultivation in what they maintained were their forests. This consciousness was preserved through stories describing how the *gora sahebs* (white men) had tricked them and taken the forests. Although by the 1980s the people were mainly cultivating fields outside the reserves, they still knew the location and names of the village sites that had been 'swallowed' by the reserves. They were known as the juni gaothan (old villages). Although in the past the Bhils had mainly asserted the right to cultivate these lands, the movement that emerged in 1989 included the Bhils, Konkanas, and other adivasi groups, such as the Varlis. It was informed by a strong belief that whatever their community, subaltern groups in the Dangs had common interests.

In early 1991, members of the Adivasi Bhumihin Kisan Hak Rakshak Samiti went into a thickly forested part of the reserve, which they said was a juni gaothan, and began clearing it. They demanded that their right to the spot be recognized. The forest department responded by sending in the police, and about ninety people were arrested and beaten. To intimidate others, raids were carried out in their original villages. In one case, the villagers fought back, surrounding a police party and forcing the sub-divisional forest officer, to eat grass. They told him, 'You often call us cattle. Now see how it feels to be like that yourself.'[40] The police returned soon after armed with guns. They were met by a large group of Dangis armed with bows and arrows and slings. When the police acted high-handedly, they advanced menacingly, waving their weapons. A stone was hurled from a sling, and, panicking, the police retreated in confusion, abandoning two rifles.

The official in overall charge of the south Gujarat forests, R.S. Pathan, felt that a more imaginative approach was needed to counter what he saw as the 'Naxalite threat'. By now the conflict was generating adverse publicity for the forest administration outside the Dangs. Pathan had already been experimenting in parts of Surat District with new approaches to forest management, notably the formation of Forest Protection Committees (FPCs) in which local people were encouraged to be responsibile for forest protection and were given, in return, certain rights to forest produce.[41] He had achieved some success in this, and felt that the strategy could be extended to the Dangs. Ironically, the Adivasi Bhumihin Kisan Hak Rakshak Samiti had, as part of its work, already begun organizing forest protection along such lines in two villages, with guards being posted at night to prevent illicit cutting by people from outside the Dangs. Pathan did not, however, try to build on this base.

Another problem with the FPC idea was that it ran against the interests of the leaders of the forest labour co-operatives. For them, the major problem was not that of forest protection but of labour employment — preferably under their control. Through their political influence they ensured that the scheme, which was eventually unveiled in October 1991, favoured their interests rather than those of forestry. Funds were to be made available to finance a new initiative, known as *van sathe vikas* (development with the forest). Local committees were to be established to organize road-building, small-dam construction, well digging, land levelling, soil conservation, gully plugging, and to raise teak and fruit tree saplings. People were to be given bamboo at reduced rates to make bamboo items, which were to be marketed through a co-operative. The committees were to be under the village panchayats.[42] This meant that they would be dominated by the leaders of the forest labour co-operatives and their allies in the villages, and that by handing out employment they would be able to restore the hold they had lost after the moratorium on forest cutting. The whole exercise — as established in practice — had almost nothing in common with the FPCs established by R.S. Pathan elsewhere in south Gujarat.

The forest officials based in the Dangs, as well as the civil authorities, saw this as merely a distraction from the chief need, which was to restore their lost authority. They were troubled in particular that one of the leaders had Naxalite connections, for this was the first time that any militant communist had managed to establish a firm rural base among a subaltern group in Gujarat State. The officials therefore called in more than a thousand special reserve police to attack the villages in which resistance had been the greatest.

The reign of terror began in November 1991, when police officers went to Kosimda village and fired on a group of adivasis, killing a twenty-four-year-old Bhil woman named Taraben Pawar. Soon after, the authorities arrested and imprisoned the two leaders, Irfan Engineer and Virsinhbhai Patel. Local adivasi activists were seized and beaten severely. Villages were raided, people were thrashed, women molested, household utensils and grain-holders smashed, and property seized. The police threatened that if people did not leave the 'Naxalite group' they would suffer more of the same, and that Taraben would not be the only one killed. The terror continued until March 1992, effectively shattering the movement.[43]

The power of the forest department was thus reasserted. The grievances of the Dangs people remained, however, largely un-redressed. Although certain forest officers, notably R.S. Pathan, wanted to evolve new strategies, they were unlikely to succeed in the prevailing antagonistic climate against the forest department. As we have seen, this antagonism has had a long history, and is likely to continue so long as no attempt is made to work with the organizations in which the people of the Dangs themselves have confidence. Although the Adivasi Bhumihin Kisan Hak Rakshak Samiti has been disbanded, the Gram Vikas Mandali is still active. Its members could be receptive to a more conciliatory approach at this juncture. What has to be worked out is a system in which farmers and foresters agree to allow use of forest produce and to maintain the forest in a sustainable manner. Besides building on the past methods of cultivation, new techniques would have to be evolved to allow for a more intensive use of resources, for the population pressure is far greater today. Ideally, such a system would be extended to the current reserved areas. This might, in fact, provide the best means for preserving such tracts, which are now threatened by illegal cutting by people from outside the Dangs.

Even if the forest department radically rethinks its basic approach, which seems unlikely, there would still be the problem of alternative employment; unless this is provided, the pressures on the forest from farming will continue to grow. However, the whole issue of alternative employment is highly politicized. Dangi politicians want to control jobs, as this is a critical element in building a support base. Until 1986 this was done through control over the forest labour co-operatives, which had become little fiefdoms. Now, they hope to use the van sathe vikas initiative to this effect. Much of the money generated by such activities, whether cooperative or committee-based, will almost certainly be misappropriated by the politicians. The benefits for the majority of the people will be minimal. Such abuses of power can be fought only politically, and, again, bodies such as the Gram Vikas Mandali provide most hope in this respect. Assistance and funds need to be provided to bodies with a record of working in an egalitarian way, to establish more equitable forms of development.

The history of the Dangs over the past two centuries has been one of defeat and loss of land and livelihood. However, it has also been a history of resistance by the people of the Dangs to the idea that farming in any way represents a threat to the forest. Despite often harsh repression, this belief prevails to this day. As yet, the Gujarat government has refused to accept that the people might have a case. Until it does, the contradiction between forest use and forestry will remain, the forests of the Dangs will dwindle year by year, and the people will continue to live in poverty and need.

Notes

Some parts of this essay have appeared in my contribution, in David Arnold and David Hardiman (eds), *Subaltern Studies VIII: Essays in Honour of Ranajit Guha* (New Delhi, 1993). Abbreviations used in the notes are as follows: BA: Bombay Archives (Maharashtra State Archives, Bombay); BRO: Baroda Records Office; F.D.: Foreign Department; NAI: National Archives of India, New Delhi; P.D.: Political Department; P. & S.: Political and Secret; Pol.: Political; R.D.: Revenue Department.

1: Shiney Varghese, 'Women, Resistance, and Development: A Case Study from Dangs, India', *Development in Practice* 3(1): 11, 1993.
2. *Census of the Bombay Presidency*, 1872, pt. 3 (Bombay, 1875), p. 632.
3. R.E. Enthoven, *The Tribes and Castes of Bombay*, vol. 2 (Bombay, 1883), p. 188.

4. A. Lucas to collector of Khandesh, 1 September 1891, BA, R.D. 1892, 144/948. This report says that plough cultivation was confined almost entirely to Konkanas — hardly any Bhils or Varlis had ploughs. An account of a raid by Bhils on a village in the plains of Gujarat in 1809 mentions that nineteen iron ploughs were part of the booty carried away. It is possible that the Bhil chiefs encouraged such cultivation by the Konkanas by providing them with implements in this way. A. Skaria, 'A Forest Polity in Western India: The Dangs 1800s–1920s' (Ph.D. diss., University of Cambridge, 1992), p. 66.

5. Some Konkanas also carried on this form of agriculture. Kumri and dalhi were generic terms for it: the Dangi term was *khandad*. A. Skaria, 'A Forest Polity in Western India: The Dangs 1800s–1920s' (Ph.D. diss., University of Cambridge, 1992), p. 160.

6. This form of cultivation is described by A.T. Shuttleworth, 6 April 1871, BA, R.D. 1871, 18/552; report on Dangs by E.M. Hodgson, 29 August 1902, BA, R.D. 1902, 107/949 II; Shuttleworth to E.W. Ravenscroft, 27 August 1877, BA, R.D. 1877, 69/1566.

7. Shuttleworth to Ravenscroft, 27 August 1877, BA, R.D. 1877, 69/1566.

8. Inquiry by J. Campbell, 5 July 1871, BRO, Rajdaftar office, serial no. 710, *daftar* V/426, file 113, Residency file no. 167A and 316, pp. 126–38.

9. In 1828 the Gaikwad paid Rs 910 as protection money to Silpat. Silpat received almost no other such payments. G. Giberne to J. Williams, 20 January 1828, BA, P. & S. Dept. 1828, 29/320.

10. W.J. Morris to H.W. Reeves, 22 May 1843, NAI, F.D. Pol., 24 February 1844, 1–10.

11. Report by W.J. Fenner on a tour of the Dangs in April 1856, in A. Gibson, *Forest Reports of the Bombay Presidency for the Years 1849–50 to 1855–56* (Bombay, 1857), p. 80.

12. Memo by the Gaikwad of Baroda, August 1864, BRO, Rajdaftar office, serial no. 719, daftar V/479, file GRS/7, Residency file no. I–73, p. 177.

13. Digby Davis, 14 September 1888, *Annual Reports on Western Bhil Agency, Khandesh*, British Museum, I.S. BO 1/2.

14. Petition by Anandrao Virsingh, *Raja* of Vasurna, to Government of Bombay, 15 March 1843, BA, P.D. 1843, 20/1441.

15. W.H. Propert to E.P. Robertson, 26 March 1878, BA, R.D. 1878, 58/127.

16. For such a calculation, see Bombay Forest Report 1863–64, in *Forest Reports of the Bombay Presidency for the Years 1860–61 to 1867–68* (Bombay, 1869), pp. 115–6.

17. R.H. Madan to A.T. Shuttleworth, 28 October 1891, BA, R.D. 1892, 144/948.

18. *Administration Report of the Forest Department in the Bombay Presidency, including Sind, for the Year 1888–89* (Bombay, 1890), p. 13.

19. D. Worster, *Nature's Economy*, p. 127.

20. Ibid., p. 171.

21. Jack Westoby, *Introduction to World Forestry: People and Their Trees* (Oxford: Oxford University Press, 1989), p. 80.

22. Richard Tucker, 'Forest Management and Imperial Politics: Thana District, Bombay 1823–1887', *The Indian Economic and Social History Review* 16(3): 281, July–September 1979.

23. Ramachandra Guha, 'Forestry in British and Post-British India: A Historical Analysis', *Economic and Political Weekly* 18(44): 1884, 29 October 1983.

24. Report of the Commission on Dang Forests, September 1878, BA, R.D. 1879, 90/947.

25. Minutes by Sir Richard Temple, 7 July 1879.

26. Agreement with Dang chiefs at Pimplaidevi, 2 May 1889.

27. Bombay Government Resolution, 15 December 1893, BA, R.D. 1893, 133/948; report by A. Cumine, 26 December 1894, BA, R.D. 1895, 314/948.

28. Report by J.A. McIver, 12 July 1899, BA, R.D. 1901, 151/949 I.

29. Report on Dangs by E. M. Hodgson, 29 August 1902, BA, R.D. 1902, 107/949 II; Annual Administration Report on Dangs, 1902–03, BA, Education Departmemt 1903, 64/739; report by G.E. Marjoribanks, 3 July 1911, BA, R.D. 1911, 120/1113.

30. *Administration Report of the Forest Department in the Bombay Presidency, including Sind, for the Year 1903–04* (Bombay, 1905), p. 7.

31. Annual Administration Report on Dangs, 1906–7, BA, Education Department 1908, 63/739.

32. This account of the 1911 outbreak is based on reports by G.E. Marjoribanks of 16 May, 2 June, and 3 July 1911, and a report by F.G.H. Anderson of 3 August 1911, BA, R.D. 1911, 120/1113.

33. Ghanshyam Shah, 'Growth of Group Identity among the Adivasis of Dangs, Gujarat', *Journal of the Gujarat Research Society* 34:1, April 1972.

34. Ajay Skaria, 'A Forest Polity in Western India: The Dangs 1800s–1920s' (Ph.D. diss., University of Cambridge, 1992), p. 311. Although princely states were abolished in India after Independence and the Bhil chiefs ceased to be considered rulers, they continued to receive their annual handout, which in 1954 was reclassified as a 'political pension'. They are given this at an annual function held at the time of Holi at Ahwa.

35. Satyakam Joshi, 'Dakshin Gujaratni Jangal Kamdar Sakhari Mandalio; Ek Abhyas' (Ph.D. thesis, South Gujarat University, 1992), pp. 413–22.

36. Shiney Varghese, 'Women, Resistance and Development', *Development in Practice*, p. 5.

37. I have made these calculations using figures in *District Gazetteer: Dangs*, p. 211, and 'Dangna Adivasioni Ladat: Ek Tapas', *Arthat* 10(2): 24, April–June 1991.

38. S.P. Punalekar, 'Agricultural Profile of Dangs District', mimeograhed report (Surat: Centre for Social Studies, 1989), p. 1.

39. See Shiney Varghese in *Development in Practice*, p. 7.

40. Ibid., pp. 11 (footnote 16) and 14.

41. R.S. Pathan, N.J. Aral, M. Poffenberger, *Forest Protection Committees in Gujarat: Joint Management Initiative* (New Delhi, 1990).
42. Interview with N.K. Sinha, district forest officer, Dangs, December 1992.
43. Irfan Engineer, 'Dang: Report of Struggle and Repression of the Tribals', *Mainstream,* June 1992. Irfan Engineer and Virsinhbhai Patel were released at the end of March but have been banned from entering the Dangs.

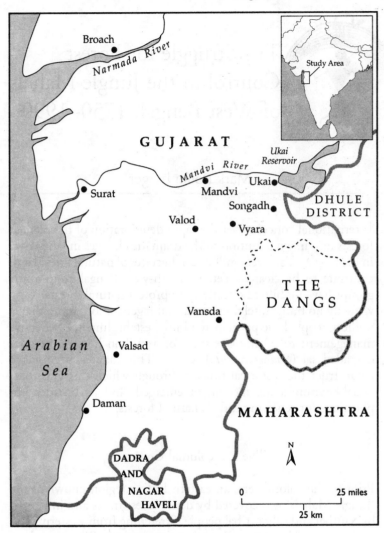

FIGURE 4.1 The Dangs

Chapter 5

The Struggle for Forest Control in the Jungle Mahals of West Bengal, 1750–1990

Mark Poffenberger

International concern over the rapid deterioration of the planet's forests has drawn attention to the dramatic changes in land cover in the tropics. Each year millions of hectares of natural forest lands continue to be cleared. Yet, in south-west Bengal community groups have mobilized effectively to protect natural forests. With virtually no budget, relying on natural regeneration, more than a million people have participated in the establishment of effective management for nearly one-third of a million hectares of once degraded sal (*Shorea robusta*) forests. This chapter attempts to reconstruct the historical process through which this grassroots social-environmental movement emerged, and its broader implications for renewing India's natural forests.

The Pre-colonial Context

Prior to the colonial era, an eastern Indian region known as the 'Jungle Mahals' was covered by dense forest tracts and inhabited by Santhal and other tribal people. Stretching from western Midnapore in south-western Bengal across southern Bihar, much of the area was wild and remote, yet it provided well-stocked forests for hunting and gathering, and small river plains for agriculture (see Figure 5.1). The hilly, rugged Jungle Mahals presented an ideal environment for tribal groups to live in, avoiding the oppression of overbearing rulers. This chapter outlines strategies used by tribal communities to maintain their independent control over land and forest resources. It focuses on the western part of the

Jungle Mahals including the police stations (*Thana*) of Garbetta, Binpur, Gopiballavpur, Salboni, Silda, and Jhargram in western Midnapore District. The area is primarily populated by Santhal, Bhumij, and Mahato tribals, and some low-caste Hindus.[1]

In the pre-colonial period, the Jungle Mahals were nominally under Mughal control. Due to the inaccessibility of the area, however, little attempt was made by outsiders to extract revenues or exert political authority. Mukerji comments, 'Although the Muslim Jagirdars were posted in Rajnagar in Birbhum, it seems a large number of tribal folk remained insular in the hilly regions.'[2] The Santhal and Bhumij tribal communities inhabiting the forest practiced shifting (swidden) cultivation, as well as hunting and gathering forest products.[3] Wild fruits, roots, herbs, and the nutritious flowers and fruit pulp of the mahua tree provided much of the diet, making them less dependent on agriculture and highly mobile. Tribal villages were also actively engaged in trade in firewood, silk, resin, deer and buffalo horns, wax, honey, bark fabrics, lac, medicines, and charcoal.[4]

Hamilton, writing in the 1820s, notes that when the forest dwellers encountered the 'least oppression' from rulers or locally powerful groups, they fled.[5] The forest-dwelling communities of the Jungle Mahals could also resist incursions into their areas. Their superior knowledge of the jungle and their hunting skills made them an effective guerrilla force. Pre-emptory raids on lowland groups expanding into the forest areas also provided economic benefits. Some Bhumij communities gained the reputation as *chaurs* (robbers) from their aggressive raids on the plains. Rather than enter into conflicts with the jungle people, many local rajas (rulers) and zamindars (landlords) preferred to leave them alone and not attempt to extract taxes from them.

Tribal communities that maintained forest-oriented, self-sufficient economies were best able to thwart outside political domination. They alternatively protected their political autonomy and forest resources through warfare and withdrawal. Where tribal communities had grown more dependent on settled agriculture, local zamindars made agreements with them, giving them formal land tax exemptions if they would serve as *paiks* (militiamen).[6] J.C. Price notes, 'The aborigines of the jungle lands had been granted *paikan* lands (free of taxes) by their Rajas for their subsistence, and they have been enjoying these lands on hereditary basis for long periods in lieu of their services of police duties to the jungle-Raja.'[7] To protect against chaur raids, some zamindars and local rajas made *ghatwal* grants to local chiefs if they would guard mountain passes. Baden-Powell noted, 'The chief of the frontier region (*ghatwal*) was allowed to take the revenue of a hill or forest tract on the condition of maintaining a police or military force (*paiks*), to keep the peace and prevent raids of robbers on to the plain country below.'[8]

   • The polity of the Santhal communities of south-west Bengal and other tribal forest-dwelling groups, was based on either community councils or village chiefs. In the former case, the village is governed by a council of elders or panchayat whose membership is decided annually by community members. The traditional village council operates under a *pargana* (council) comprised of ten-to twelve village panchayats, while the final authority rests with the *Lo Bir* (forest council), which may extend over an entire district and is the final court for dispute arbitration. The annual hunt organized by the Lo Bir provides the basis for inter-village political organizing, conflict resolution, military organizing, and joint decision-making. Duyker reports, 'The social significance of the Lo Bir far outweighs its economic importance.'[9] More specifically, the Lo Bir appears to have provided a unifying mechanism among dispersed Santhal communities, both over space and time. The council provided communication channels through which information regarding social and political issues could be exchanged and an organizational body through which some consensus on tribal policies might be reached.

   During the pre-colonial period, and up to the present, the belief-systems of the forest communities were strongly grounded in the worship of nature. Religious festivals are linked to both the agricultural cycle and the flowering and fruiting of the forest trees. The Santhal New Year, for example, begins with the blossoming of the sal tree in March. In tribal belief, the links between the health of the forest, fertility, and prosperity are clear in the following lines from this Baha festival song:

> When the *sal* trees are in leaf,
> On the mountain,
> How lovely they look,
> Wealth in the house . . .[10]

This subsistence-oriented, isolationist life of the tribal communities of the Jungle Mahals began to change with the emergence of the British colonial power in Bengal in the late eighteenth century.

Eroding Tribal Land and Forest Rights during the Colonial Period

In 1760, the district of Midnapore was transferred to the British by Mir Qasim, making it one of the first districts in India to be brought under British rule.[11] The area was comprised of vast tracts of forest, broken only by patches of farmland. Early English colonialists were unhappy with the lack of revenue generated and the controls imposed on the area. As Hamilton notes, 'These jungles were occupied by a poor, miserable, proscribed race of men called Santhals . . . and the land was under the dominion of chieftains who had never been reduced to submission by the Mughals and who "never paid any regular rents for their lands".'[12]

During the late eighteenth century, the British sent military expeditions into the Jungle Mahals in an attempt to extend their authority and extract land revenues. According to Richard Becher, an officer of the company writing in 1769, 'When the English received the grant of the Dewani, their first consideration seems to have been the raising of as large sums from the country as could be collected to answer the pressing demands from home and to defray the large expenses here.'[13] The forest chieftains and tribal communities resisted, ambushing British forces and harassing them whenever possible. According to one British source from the period, 'As soon as the harvest is gathered in they carry their grain to the tops of the hills, or lodge it in other fastnesses that are impregnable; so that whenever they are pursued by a superior force they retire to these places, where they are quite secure, and bid defiance to any attack that can be made against them.'[14] Local zamindars also initially resisted the imposition of colonial authority, refusing to pay their taxes, organizing their paiks to resist, and falling into arrears on their taxes. In 1798, widespread violent resistance disrupted revenue collection activities in the Midnapore area, forcing the Company to restore lands, that had been put up for sale for failure to pay taxes to hereditary chiefs.[15]

Through superior force, however, the British gradually

succeeded in extending their control to the area through the nineteenth century. As this process continued, the British empowered a new class of zamindars to control and tax local forest communities, encouraging them to open forest land for cultivation. Individual villages were established under *Mandali* tenure which could be incorporated into the revenue collection system. The *Mandal* (village chief) brought tribal labourers with him to convert forest to agricultural land. The zamindar financed the migration and subsistence of the tribal community (usually Santhals) until the land became productive. Some of these zamindars were allocated huge tracts of land. For example, 'The Pargana Cundar was one of the largest Zamindaries of Midnapore . . . containing about 663 villages and over 130,000 bighas of land.'[16]

The tribal communities of the Jungle Mahals resisted the imposition of the taxation system through a series of armed revolts. The first, referred to as the *Chaur* Rebellions,[17] lasted from 1767 to 1800. Tribal guerrillas were so effective that 'even as late as 1800, after nearly forty years of British occupation, a collector reported that two thirds of Midnapore consisted of jungle, the greater part of which was inaccessible.'[18] Yet, gradually the East India Company succeeded in strengthening its control, despite subsequent revolts by the forest people, such as the Naik Revolt (1806–16). By early nineteenth century, while courts, jails, and the district police were ineffective, collection of land revenue was becoming a routine matter. Under the Permanent Settlement Act, by 1866, 1369 *zamindari* estates had been established in Midnapore, giving absolute ownership of agricultural lands and forest tracts to the zamindars as long as they paid government taxes.[19] In order to meet their tax obligations, zamindars were anxious to bring in tribal and peasant cultivators to clear forest land and convert it to agricultural land. Tribal communities often lost control of their paikan lands to zamindars and the Company, becoming tenant farmers. Due to the greater farming expertise of Hindu peasant cultivators, tribals were often displaced by zamindars in favour of the former, further exacerbating their social and environmental dislocation.

Tribal and low-caste families also suffered at the hands of moneylenders (*mahajans*). The *diku* (plains people) moneylenders who migrated into the Jungle Mahals began to displace the zamindars as a source of credit for small farmers. The mahajans were far more effective than the zamindars in converting outstanding loans into land mortgages and then foreclosing on them when the borrower failed to pay. Moneylenders often charged an exorbitant interest rate of fifty per cent per year, forcing defaulters to migrate or become tenant farmers. Dasgupta cites McAlpin who wrote in the earlier twentieth century that many tribal and low-caste communities lost the majority of their farmlands between 1892 and 1906. Of the 120 Santhal villages that had owned land which McAlpin surveyed, '35 had sold their rights to pay off debts, 6 surrendered their rights, 19 had their rights forcibly taken away, and for 54 the process of loss was unknown.'[20] As Baden-Powell writes, 'The Mahajan entrenched himself in the rural economy which came to be dominated by him.'[21] With the elimination of tribal landowners, the Bengali mahajan landlords and the large zamindari companies came to control land resources, raising farmland rents drastically and eliminating many of the forest-use rights previously enjoyed by tribal and low-caste communities.

The process of clearing forests for agricultural land had sweeping ecological implications, especially for the river systems and soil conditions. Removal of the forest cover allowed torrential monsoon rains to wash away the shallow topsoil, leaving an exposed laterite hard pan that made farming virtually impossible in many areas. As the forest was cleared, traditional forest-based industries like tussar silk declined dramatically, as did the population densities in Chandrakona, Ghatal, and other regions of the Jungle Mahals.[22]

By the 1860s, pressure on the forests of the Jungle Mahals grew further as the growing railway system demanded immense quantities of sal logs to provide 'sleepers' for the rail bed. The construction and opening of the Ajay-Sainthia and Sainthia-Tinpahar railway lines in 1860, followed by the construction of the main line of the Bengal-Nagpur railway in 1898, stimulated commercial felling. Commercial demand for timber raised the value of forest lands.[23] Timber merchants rushed in, even before the railway lines opened and began leasing or purchasing large tracts from the Midnapore Zamindari Company and other zamindars.[24] As the value of the forest increased, leaseholders and zamindars began imposing strict controls on forest use by local communities, restricting or eliminating traditional usufruct rights enjoyed under the Mandali land-tenure system. When tribals and low-caste groups appealed to the Settlement Department, their complaints were usually dismissed, on the grounds that the 'encroachments of the landlords were justified by "unavoidable economic circumstances".' Dasgupta convincingly argues, 'It was with the active connivance and supervision of the imperial bureaucracy that the destruction of the traditional jungle rights was carried out.'[25] Tribal communities were charged fees to gather fuelwood or cut roofing poles. Zamindars sold lease rights for the collection of lac and cultivation of silk. They also carried out periodic searches of forest communities to detect illegally cut fuelwood or timber. If forest-products were found, family members would be beaten and sometimes killed. Bradely-Brit notes that in Chotanagpur new zamindars also demanded that tribals pay taxes on mahua flowers, and sometimes cut down the trees and sold them for timber if they failed to pay.[26] As customary access to the forests was restricted, friction grew between tribal and low-caste communities and the local zamindars.

In response to their growing marginalization, in early 1855 six to seven thousand Santhal tribals from Birbhum, Bankura, Chotanagpur, and Hazaribagh began meeting to organize a resistance. Under the messianic leadership of four Santhal brothers, on 16 July 1855 some ten thousand tribal rebels 'stood their ground firmly and fought with bows and arrows and a kind of battle axe'[27] in a battle near Pirpaiti. Eventually, the revolt collapsed after half their number were reportedly killed. Despite their defeat, the *Hul* Rebellion (as it is known among the Santhals) 'profoundly influenced the ideological development of many Santhal communities',[28] and lives on in the songs and oral traditions of the tribal people of the Jungle Mahals.

Towards the end of the nineteenth century, many estates of small landowners were absorbed by large landowners. These included the British-held Midnapore Zamindari Company (MZC), the Jhargram Raj, and the estate of the Raja of Mayurbhanj. Throughout the latter part of the nineteenth and first half of the twentieth century, many forest communities in the Jungle Mahals became increasingly indebted to moneylenders and tax collectors, causing widespread mortgaging and loss of their agricultural lands. McAlpin, writing in 1909, notes a 'general transfer of Santhal lands to non-Santhals', observing that most sales of Santhal lands were due to previous debts' and that the land was generally sold for as little as Rs 10 per *bigha*.[29] A report in 1947 revealed that the average agricultural landholding size of tribals had decreased to 0.5 acre of owned land and 1.2 acres of sharecropped land, highlighting the precarious nature of such marginal farming operations.[30] The Bengal Tenancy Act and the Zamindari Abolition Act of 1953, which were designed to assist tribal and disadvantaged farmers, failed to have the intended effect. In some cases, the acts even facilitated the eviction of sharecroppers. By the early 1970s, many tribals and low-caste people in Midnapore had been reduced to agricultural labourers and sharecroppers.

The alienation of private lands was an important element in the impoverishment of tribal and low-caste communities; so too, was the loss of income in cash and kind from forest-based activities, as the forests were cleared. Writing in the mid-1850s, Sherwill noted that nearly 20 per cent of the population of Birbhum was involved in the collection, processing, weaving, and marketing of tussar silk.[31] By the early 1970s, the weaving industry had significantly declined, partly due to the need to import cocoons, which had previously been available in abundance in the area.

Government attempts to extract greater revenues and impose increased controls over freshwater fisheries throughout the latter half of the nineteenth and early twentieth centuries also restricted access of marginal communities over once common property resources.[32] In response to these restrictions, Santhals in Midnapur and Bankura carried out mass loots on fishponds in 1922 and 1923. 'In April 1923 for instance there was a wave of looting of fishponds and violation of forest rights over an area of 200 square miles extending from Jamboni and Gopiballabhpur westward to Ghatsila in Singhbhum district of Bihar and northwards through Silda . . . to Bankura district'. Sarkar quotes an official of the time who noted that the Santhal 'will tell you how in his father's time all jungles were free, all *bandhs* (ponds) open to the public' and that this action was simply 'carrying on an old tradition', an attempt to bring back the 'golden age'.[33]

Changes under the Populist Government

A major shift in the erosion of land-control rights of disempowered people in the Jungle Mahals began in 1967 when the newly-elected United Front Government announced its intentions 'to distribute surplus land among the landless and halt the eviction of sharecroppers'.[34] Despite a basic agreement on the need for land reform among the fourteen member-parties of the United Front Government, the coalition was unable to develop effective implementation programmes for such policies. Policy implementation was also resisted by the more conservative party members, and by the need to comply with bureaucratic and judicial procedures. Nonetheless, the announcement of land reform plans stimulated widespread interest in rural areas. Dispossessed groups, particularly the Santhals, began to organize. They met in groups, bringing their traditional weapons with them and often confronting armed police. 'Between March and May 1967, nearly one hundred incidents involving *kisans* [peasant farmers and agricultural labourers] armed with bows and arrows, occupying land and symbolically establishing their ownership by ploughing small parcels, were reported to the district police.'[35] The famous uprising at Naxalbari, in Darjeeling district, took place a few months later, when a local Communist Party of India Marxist (CPIM) cadre and a tribal leader named Jangal Santhal organized over 600 tribals to attack local government officials and landlords.

The CPIM, embarrassed by the actions of some of its members, expelled them from the party. Other disillusioned CPIM members joined those expelled, ultimately forming a new party — the CPIML, generally called the Naxalites. The Naxalites rejected the parliamentary system and saw no alternative to an armed struggle and a protracted insurgency movement. The Naxalites carried their message of land reform and class struggle to the poorest communities, finding strongest support among the Santhals.

Throughout 1969 and 1970, violence shifted to Midnapore where the Naxalites were encouraging the adoption of Maoist-oriented politics, similar to those expressed in Naxalbari. The emerging Naxalite leadership effectively enlisted the support of disenchanted tribals and landless labourers. The houses of landlords were raided and stockpiled rice redistributed. Killings were also carried out. Some forest communities were also discontented with the forest department and its policy of providing elites and contractors with low-cost resource-exploitation leases. While forests were logged for timber and bamboo, villagers lost the raw materials they required for their subsistence and commercial needs. According to Pritish Dasgupta, a leader of the Midnapore uprising, the Dom tribals, in particular, were upset by the high prices contractors charged them for bamboo.[36] The high prices and fuelwood scarcity experienced by potters, blacksmiths, and other caste groups also increased antagonism toward the forest department and those who acted as contractors for them.

Many village elites and landowners fled the area and the CPIML began to institute village committees to fill the political vacuum it had created.[37] The broad-based participation was so extensive that the authorities had difficulty re-establishing control, and refrained from directly confronting huge groups of militant labourers, sharecroppers, and smallholders. Gradually, however, by mid-1970 more than 700 Naxalite leaders were rounded up. Yet, the resistance and the new political leadership that emerged had a fundamental influence on community attitudes to private and public forest land.

From Conflict to Compromise: Experiences of the West Bengal Forest Department, 1972–82

In 1970, the relationship between foresters and villagers in south-east Bengal was tense. The state forest department restricted community access, but continued to lease felling rights to timber companies in accordance with the working plans. At the same time, poor tribals began cutting the shrinking state sal forests and questioned state authority in the wake of the Naxalite uprising. On occasions when the forest department sought police assistance to restrain villagers from cutting timber and fuelwood, they were often met with forceful resistance from community groups. Violent confrontations often resulted in deaths from shooting and injuries on both sides.

In 1970, in another corner of south-west Bengal, A.K. Banerjee, an Indian forest service officer, had been appointed chief silviculturalist at a small forest-research site at Arabari. Experiments were being conducted with native sal, teak, eucalyptus, and other timber species. The trials were constantly being disrupted by villagers cutting fuelwood and grazing their cattle on the experimental plots. The silviculturalist began meeting with members of the eleven villages surrounding Arabari, 30 kilometres to the north of

Midnapore. The officer attempted to offer the villagers a comprehensive employment programme, to absorb them in plantation work. In return he asked them to stop grazing activities and cutting fuelwood on the field station. Due to limited budget and employment opportunities, he later revised the arrangement promising them a 25 per cent share of the sal timber and rights to all minor forest products including leaves, medicinal plants, fibre and fodder grasses, mushrooms, and fruits. This agreement appealed to the communities, and local villagers ceased their grazing and cutting, and began protecting the forest from use by outsiders. In 1972 the first Forest Protection Committees (FPCs) were born in the villages around the Arabari forests.

While successful examples of joint management agreements were beginning to emerge in the Arabari area during the early 1970s, throughout the decade they remained isolated cases with little effect on routine forest management systems within the state. Nonetheless, these early experiences demonstrated that opening communications with forest communities could effectively reduce conflicts between the forest department and forest user-groups. Through discussions, forest officers were able to identify terms for effective management partnerships. By formulating agreements that responded to the economic needs of forest communities, new incentives that resulted in the emergence of effective controls on forest exploitation were created.

By the early 1980s, in a growing number of communities, villagers were starting to take over public forest lands without the forest department's permission. Neighbourhoods began forming volunteer patrols. Local people who were found cutting green wood or grazing animals were warned by village volunteers. Repeat offenders from the participating villages were fined, and outsiders were turned over to the forestry field staff. Most confrontations occurred during the first and second year of protection, after which the rights of the protecting communities were generally recognized by outsiders. Where rootstock was still viable, community-based forest protection usually resulted in the rapid regeneration of degraded natural forests. Natural regrowth led to substantial increases in biomass productivity and the enhanced availability of a range of important minor forest products. The capacity of degraded natural forests to regenerate rapidly and produce fodder, fuel, fibres, and other valuable materials appears to have been instrumental in sustaining community-protection activities.

Independent community forest protection initiatives, as well as those encouraged by forest department officers, tended to be carried out in isolation from one another. Community forest department agreements were informal and generally had little validity beyond the term of the individual forest officer, who was usually rotated to a new area every three years. There was little effort to co-ordinate the terms of the agreement in one area with those being offered by forest officers in other territories. Officers exploring new agreements with communities, however, were communicating their experiences to their superior officers, and the progress they were achieving, particularly in Arabari, was attracting attention. However, it was not until the mid-1980s that some senior forest officers began to encourage field staff to pursue similar negotiations in wider areas throughout the south-western part of the state. They urged young district forest officers to encourage field staff to work with forest communities. To accelerate the formation of local community-management groups, the conservator told range and beat officers that their performance in forming Forest Protection Committees (FPCs) would be monitored and awards given to those officers who were most successful.

Through informal discussions and small group meetings, the circle conservator and his district forest officers gradually conveyed a message to field foresters that the department was committed to involving communities meaningfully in forest management. Local party and panchayat leaders who were also informed of the department's new strategy, began sending this message to communities. The several dozen FPCs existing in 1985 increased to 2092 by April 1992, covering 308,000 hectares of forest land throughout Midnapore, Bankura, and Purulia districts.[38]

The expansion of joint management activities in West Bengal, particularly during the late 1980s, was facilitated by the growth of the department's social forestry programme. These programmes required foresters to negotiate formally with village groups and brought a community-extension orientation into these agencies for the first time. At the same time, small cultivators, landless people, and tribal families found new authority as the power of landlords and contractors was diminished through political and legislative acts. The growing openness and flexibility of the state forest department, and its attempts to minimize corruption, allowed field staff to gain respect in the eyes of the communities. The new joint management initiative promoted by some senior officers urged field staff to stress the environmental problems caused by deforestation and the benefits of sound management, rather than raise expectations of the communities with promises of guaranteed forest employment and timber revenues.[39]

While the officers of the West Bengal Forest Department (WBFD) feel justifiably pleased by community concerns over forest degradation and the willingness of village members to mobilize to protect natural forests, the extraordinary increase in the number of FPCs indicates that a receptive social climate was already present prior to the initiation of programme expansion. Further, many communities in neighbouring Bihar and Orissa were also establishing forest management groups, but without similar forest department programmes. While the WBFD supported the formalization and expansion of these organizations, the driving force behind this shift in the tenure of public forest lands appears to have emanated from the community, the concerns of its members, and their shared history. The historical events outlined earlier in this essay suggest that communities in the area have mobilized repeatedly over the past 200 years to protect their resource rights from manipulation by outside groups (see Figure 5.2). To that extent, the emergence of community resource management groups reflects another attempt by groups with low socio-economic status to reclaim control over their environment. The process of FPC formation, the extent to which it was driven by a community concerned by environmental changes and the encouragement by field staff, are reflected in the following case studies.

The Emergence of Forest Protection Committees in Chingra Forest

The Chingra Mouza forest in Midnapore district in south-west Bengal is located on the Bihar border (see Figure 5.3). A century ago the *Munda* tribal community of Chinga was surrounded by dense sal forests. The trees were of great girth, three to four feet in diameter. The region was sparsely settled and there was little pressure on the forest. Occasionally trees were felled to construct roof frames, make ploughs and axe handles, and meet other domestic needs; however, this had a negligible effect on forest density. The villagers collected many products from the forest. The forest was particularly important as a source of supplementary food.

Over the past century, the original sal forests of Chingra receded. Throughout the 1960s and 1970s the forest department leased cutting rights to contractors who used village labour to log the area. The growing rural population of the area also found fuelwood-cutting a convenient way to generate additional household income. During the droughts of 1981–82 the monsoon rice crops failed. Desperate villagers from communities throughout the area attacked the remaining sal forests. The increasingly denuded forests were used as pasture lands. The cattle and goats consumed or trampled the young sal shoots as they emerged. The sal was no longer able to regenerate, and the root systems began to die off. The exposed forest floor gradually lost its topsoil through water and wind erosion.

In Chingra, the local forest officer was unable to control the vast numbers of people and animals exploiting the 450-hectare forest. By 1983, he had given up all hope of saving the natural forest from further degeneration. The forest officers decided to establish a small eucalyptus plantation on some of the degraded forest lands near Chingra village. However, when the healthy saplings reached a few feet in height, people from a village near Chingra began cutting them down for fuel. Seeing this, the Chingra villagers thought that they should also benefit from the government project and cut down the remaining trees. The local forester was further frustrated by the failure of the plantation project.

Meanwhile, the young men of Chingra were aware of the deteriorating state of their environment. They had often heard stories from their elders about the beautiful forests that had once surrounded the community, and the many things these forests produced. An old Munda tribal man said, 'The forests are like your eyes; you don't realize their value until they are lost.' A youth named Mahadev Munda Singh was particularly disappointed over the failure of the eucalyptus plantation. Mahadev had finished his high-school studies and understood the ecological importance of the forest. He was also a popular boy in the village and was proud of his community and his tribal heritage. Because his family was poor, he worked as a labourer shovelling sand. One day in 1984 he approached the local forester and asked if he could re-establish the small eucalyptus plantation. The forester responded by allocating a small amount of departmental funds with which he started a nursery. His friends helped him plant the seedlings. He encouraged ten members of the village youth club to join him in protecting the young trees. As the trees grew, the youth saw the sal trees and other plants in the forest also grow rapidly.

The group agreed that if protected, the entire forest could regenerate. The club asked the forester to place an additional 50 hectares of natural sal forest under their protection. They asked for a 50 per cent share of the produce to support their activities, and the forester agreed. The sal, too, began to regenerate. The group gradually extended their protection activities to the entire 450 hectares of reserve forest neighbouring their village. When people from surrounding communities came to cut firewood, the boys would try to explain to them the need to protect the forest and how it could better meet their requirements if they allowed the forest to regain its health. Sometimes, when fuelwood gatherers refused to cease their cutting, the boys would impound their bicycles. They convinced their own families not to let their cattle graze in the protected forest and chased away cattle from other villages. After three months, the regenerating young sal shoots had grown to four feet or more in height and were above the reach of the small cattle.

Seeing the success of the Chingra group, people from other villages began thinking of protecting forests near their own communities. Observing the growth of local interest in forest protection, the local forest beat officer began holding meetings with villagers to encourage them. At each meeting he invited Mahadev and his friends to talk about their activities and their hopes for forest regeneration. By 1988, nine communities on the West Bengal side of Chingra were protecting the forest, and the sal had reached a height of fifteen to twenty feet. A dense undergrowth of climbing vines, shrubs, grasses, and small palms also emerged.

The rapidly regenerating forest began to attract fuelwood

cutters from the neighbouring state of Bihar. One night a band of Bihari villagers came with their axes, bicycles, and bullock carts to fell the Chingra forest. The Biharis, attempting this 'mass loot' of the forest, were soon confronted by Mahadev and his club members, as well as men from many of the neighbouring villages. Mahadev said, 'We took our spears, arrows, and axes and faced them eyeball-to-eyeball. We talked and told them the forest was protected now and that the trees were no longer available for commercial cutting.' In the end, the Chingra villagers offered to give the Biharis dead twigs and sticks to meet their subsistence fuelwood needs. Later, the Biharis organized their own groups to protect forests near their village on the Bihar side of the border.

Many intervillage meetings have been held over the past three years to work out agreements, settle disputes, repulse outside users, determine territorial protection responsibilities, and establish usufruct rights among participating communities. While eight villages have joined Chingra in protecting the local forests, the community of Talgram has not co-operated. This village is comprised of immigrants from Orissa who were brought into the area by a local landowner as agricultural labourers. On 18 November 1990 a special meeting was called by Mahadev and his friends to bring all the communities including the Orissa migrants together, to invite them to join the management group. The panchayat leaders and range and forest beat officers were also invited. Mahadev and his friends hoped that this would result in a common understanding regarding management priorities, a clarification of community usufruct rights, and the authorization of the community to use fines and physical force when necessary, to protect the forest from outside users.

The Chingra case reflects the decentralized nature of FPC group formation. It also illustrates the way in which village leaders like Mahadev were able to work with field staff and other neighbouring communities to identify forest areas for protection and reach agreements. The ability of local communities to take the lead in defining management territories was the key to the success of the programme. While the field staff helped facilitate this process by encouraging group meetings and authorizing community protection activities, successful FPCs frequently took the initiative in organizing themselves and establishing operational controls over forest access.

The Emergence of Forest Protection in Chandana

Chandana and Harinakuri villages are located approximately 20 kilometres south of Kharagpur, in the state of West Bengal. A two-kilometre-long dirt track off the main road, after crossing rain-fed rice fields and passing through regenerating forest lands, leads to Chandana village. Another kilometre down the road, bordering the southern extension of the forest, is Harinakuri village. The forest lands in the Chandana area, totalling 160 hectares, are surrounded by Chandana and Harinakuri villages on the south and by Nidata and Babunmara villages to the north (see Figure 5.4). The land slopes gradually downwards as it drains into the Kele Ghai river to the north of the four villages.

Most of the villages in the area are inhabited by low-income scheduled castes, tribals, and farming families. In Chandana village there are thirty-eight households, half of whom are Bhumi tribals and the rest scheduled castes. In Harinakuri there are thirty-one families, primarily of the Naik scheduled caste. The Naiks claim to have worked as mercenaries for a local raja until about a hundred years ago when they moved into this forest area which was being opened for agriculture by a large landowner. Since settling in the area, most of the villagers worked as agricultural labourers and tenant farmers until the state land reform programme granted them titles for local rainfed rice lands. Historically, these communities have depended on the neighbouring forest lands for fuel, fodder, supplemental food, medicines, and fibers.

Forest Management History

According to Lokhun Sahu, a 65-year-old Chandana villager, the surrounding forest was once comprised primarily of first growth sal trees. During the years of British colonial rule, the forest tracts of Chandana were controlled by a zamindar named Bhuwan Chandra Pal, who lived 20 kilometres away in Hundla, near Narayangar. To pay his taxes to the British Raj, the raja would periodically sell tracts of jungle to contractors for logging. After the contractor finished logging the concession area, the sal would send up coppice growth and the forest would re-establish itself. Older trees, including sal, mahua, and cashew, were left as sources of seed and fruit. During the felling, local villagers could purchase the lops and tops for fuelwood at the rate of Rs 1–2 per cart load. The raja did not allow villagers to cut poles or logs, and had guards to protect the forest against local users. Periodically, the raja would send his men into the village to see if the villagers had hidden poles or timber. If caught, the guards would beat, sometimes to death, anyone found to have stolen wood.

During the early years after Independence, little changed in forest management practice, with the zamindar continuing to control the forest of Chandana. In the early 1950s, however, the Zamindari Abolition law was passed giving the West Bengal Forest Department (WBFD) an opportunity to establish direct control over the forest lands of the south-western part of the state. Seeing that he was about to lose control of the forest, the local zamindar sold the entire forest tract to contractors who felled the area, leaving only a few fruit trees. For the next six months, communities faced a severe shortage of fuelwood; however, as the coppice growth emerged, the forest resource supply also began to recover.

From the mid-1950s through the 1960s, the WBFD exerted control over the forests of Chandana. Throughout this period, the WBFD continued the practices of the zamindars by leasing cutting rights to contractors. Consequently, the sal trees were cut every ten to fifteen years, regenerating through coppice growth. The local field officer complained that the contractors would often cut the older sal and fruit trees, a practice that was officially banned, as these trees were sources for natural regeneration. When the forest guard or the villagers attempted to stop the contractors, they were threatened. The contractors were also reported to have enjoyed political patronage, so the field staff and villagers could do little to stop them.

In the early 1970s, according to Lokhun Sahu, political organizers began visiting the community. They told the villagers that the forest was community property. In retrospect, Lokhun felt that 'the political leaders mislead us to gain our political support.' No control system existed and the villagers began indiscriminately cutting and selling the trees. By the early 1980s, the sal forests were badly degraded. In some areas, even the root systems had been extracted for fuelwood. Lokhun reported that the temperature seemed to become hotter, while rainfall diminished, and the earth became drier. He felt that the cooling breezes had ceased to blow. The villagers had difficulty finding wood for their spade handles, ploughs, and other agricultural implements. The village ponds and well dried up, making the villagers rely on the water in the river two kilometres away. The forest had been so thoroughly cut that there were no standing trees outside the village proper, and it is said that one could see all the way to the river and beyond.

In 1983, Jyoti Naik, a man from the neighbouring village of Harinakuri, began visiting Chandana village to discuss forest management problems. Jyoti, a 45-year-old illiterate farmer with only two years of formal education, was convinced that some action had to be taken to reverse the process of forest degradation. In the beginning, Jyoti visited each house to talk about the problem. He told the villagers of Chandana that if they did not begin protecting the forest it would degrade to a point where even fuelwood and leaves would no longer be available. He told them they would be forest people without a forest and that their children would have no forest resources in their adulthood. Gradually he began organizing village-level meetings. By 1984, a sufficient number of Chandana villagers were ready to protect their forests. They called a meeting of the neighbouring four villages to discuss a collaborative management programme and decided that each community should be responsible for the forest area nearest to its village. Although the Chandana and Harinakuri villages began actively protecting the forest tracts near their communities, the villages to the north of the forest, Nidata and Babunmara, were less effective in controlling access, and commercial fuelwood cutting continued. Jyoti Naik and other village leaders met with local political representatives from the area and urged them to put pressure on the north-side communities to begin protection activities. However, Jyoti noted that the politicians were afraid that they would lose votes if they did so. Currently a four-village (Chandana, Harinakuri, Nidata, and Babunmara) FPC co-ordinating board exists, for which Jyoti Naik acts as chairperson.

Experiences with Protection Activities

The Chandana FPC has experienced continuing problems in controlling illegal cutting by outsiders. Women from other villages come in groups of five or six every two to three days to cut fuelwood. These women frequently come from Bhetia village across the river to the north, or from Pora and Simildanga villages in the south. When caught in the act by Chandana villagers they are asked to go elsewhere, and if necessary they are chased away with sticks. More serious a threat are the gangs of ten to twelve men who come during the night from August to October and from February to May — the slack agricultural season — to cut sal poles for commercial sale. When outside cutting groups are active, the FPC keeps one man patrolling the area on two- to three-hour shifts. Other villagers are also watchful and notify the community if cutting groups are seen approaching the area. Occasionally, the FPC apprehends groups while cutting, confiscates their axes and levies fines.

Protection experiences in the neighbouring village of Harinakuri are similar. Since the FPC was first formed in 1979, Harinakuri has worked with Chandana and Telebanga villages to deter cutting groups from nearby villages in the north and east. According to Jyoti, pressure from nearby villages is particularly high because many members of these communities depend on fuelwood headloading as their primary source of income. Often tribal and scheduled caste members of these villages are contracted by high-caste families in towns and villages and at the Soluwa Army base to cut fuelwood and timber for them. The cutting groups often band together to overcome local resistance. In response, the Harinakuri FPC patrol an area in a group of eight to ten men armed with bows and arrows and spears. Boys with grazing animals also watch and listen for the sound of the axe upon the tree, so that they can warn the FPC about the intruders. When this occurs, the men encircle the cutting group so that it can be apprehended. In these cases, the offenders are turned over to the forest department guard and later fined by the forest department.

Economic Costs of Protection

Jyoti noted that the decision to protect the degraded forest land had a significant impact on the economy of Harinakuri. Previously, Jyoti and the other villagers had also been engaged in cutting fuelwood for sale. If a number of family members were engaged in cutting, a household might collect two to three forty-to fifty-kilogram bundles each day. In 1979, that would have generated Rs 35 to 50 per day, while in 1991 it would have yielded three times as much. Further, fuelwood cutting and carrying could be done in three or four hours in the morning, leaving time for other work. In contrast to agricultural wage labour, which is only available during certain times of the year, fuelwood cutting would generate at least two to three times more income per unit of time spent. Consequently, it was a considerable sacrifice for the community to discontinue this lucrative economic activity. This decision of the villagers in Harinakuri was the outcome of their concern over the deteriorating environment, the recognition that the level of exploitation was not sustainable, and that they would have to shift occupations in any case, once the forest resources were exhausted.

The shift from fuelwood cutting and the loss of income it entailed was softened by the land reform programme of the West Bengal Communist Party government, which transferred titles for the rainfed rice land from the landlords to Jyoti and his neighbours, who had worked as tenant farmers in the past. By not having to share their harvests with the landlord, their incomes rose.

At the same time, Jyoti and his neighbours decided to begin producing puffed rice (*chira*) for the local market. They brought small stocks of unhusked grain (*dhan*), usually twenty kilograms at a time, husking and winnowing it, and roasting it under a brushwood and leaf fire. Three men working from 4 am to 5 pm, are usually able to process twenty kilograms of rice-grain worth Rs 60, into ten kilograms of chira worth Rs 240. This means that the hourly income per man from chira-making is about Rs 4.60 per hour, or Rs 60 per thirteen-hour day. This is about three times the official minimum daily wage of Rs 24.85 per day received by agricultural labourers. It may also approximate the income generated by fuelwood headloaders if they have sufficient forest resources to exploit. While Jyoti and his neighbours have been successful in finding an alternative, and at least as lucrative a source of income as fuelwood cutting, many neighbours have not been so fortunate and must suffer the lost income or continue to exploit the forest in defiance of their neighbours.

The amount of time Chandana and Harinakuri FPCs spend patrolling the forest and the value of that time in terms of opportunity costs are difficult to calculate. Clearly, much of the time is spent during the slack agricultural season — a period of high threat. No regimented, full-time patrolling system was used; rather, villagers, especially women and children, engaged in grazing, fuelwood collection, and other forest-related activities acted as an early warning system. When news of illegal activities was given, men would move into the forest for protection activities.

The villagers were clearly concerned that as the poles gained value, the threat of a 'mass loot' by groups of outside villagers would grow. They were aware that the regeneration of the forest had substantial environmental and economic benefits that would be lost temporarily if the entire area was clear-felled. The most important advantages emerging from forest regeneration have been improved groundwater infiltration and slowed run-off and the increased availability of non-timber forest products such as tubers, mushrooms, and fibre materials. FPC members also noticed that the re-establishment of standing forest near their village resulted in a large population of birds to nest in the area. Birds were important in controlling pests, which attacked their rice crop. They also felt the forest had a beneficial effect in cleansing the air of disease. They noted that when the forest was degraded, the incidence of disease increased, and they came to associate a healthy environment with a good standing forest.

Despite their success in protecting at least 100 of the 160 hectares of disturbed natural sal forest neighbouring their villages, the villagers continue to be confronted by threats from other villages in the area that depend on fuelwood cutting for a substantial part of their income. The tribal and scheduled caste people who illegally exploit these forests are driven by economic necessity and encouraged by local and urban high-income and caste groups. Until all communities neighbouring the forest can be effectively brought into the joint management programme and their economic needs met, these emerging local-management systems will remain threatened and their sustainability will continue to be questioned.

Conclusion

The case of the Jungle Mahals indicates that the emergence of new community forest management systems in south-west Bengal is historically grounded in tribal and peasant resistance movements. In many parts of rural India, pockets of disempowered people have repeatedly organized to struggle for their survival as their resource base is increasingly being captured by local elites, moneylenders, tax collectors, and the state. In the past, each time a movement collapsed or was crushed, it would re-emerge after some time. The people of the Jungle Mahals represent a classic case.

In south-west Bengal, in recent decades, grassroots leadership has been effective in mobilizing community commitment to forest protection. The emergence of tribal and scheduled caste leaders who could accomplish this under the populist government is a testimony to the broader socio-political changes that have occurred in the state over the past twenty years. Community members, clearly, are concerned about environmental degradation in their area and are willing and able to take action to respond to the challenge. That they were encouraged by a supportive West Bengal Forest Department (WBFD) programme and helpful field staff, definitely facilitated this process.

To understand how the WBFD has moved more quickly and more successfully than other departments involved in similar efforts, several explanations have been cited. There is little doubt that the socio-political context in the state has encouraged populist programmes and a responsiveness to forest community needs. A new generation of community leaders from small farming, agricultural labour, and tribal backgrounds has emerged. Further, the department's appeal to tribal communities to protect their forest resources and its willingness to empower them coincided with a growing desire among these communities to take environmental action. Finally, the West Bengal programme did not require complex registration and budgetary allocation processes for communities to take action. Instead, the programme presented communities with a simple strategy to protect the local forest and, in turn, enjoy the benefits. As each community began protection activities, it influenced the behaviour of neighbouring villages. Villagers were forced to negotiate and discuss management issues and needs with one another, without necessarily waiting for the forest department to take action. It is this community-based chain reaction or catalytic effect that is apparently the driving force behind the rapid emergence of localized access controls on state forest lands in eastern India especially in south-west Bengal. It is likely that a similar community concern over environmental degradation in other parts of India could provide the basis for a rapidly expanding rural movement to stabilize the nation's forest and water resources.

Notes

Acknowledgements: The author would like to thank a number of persons who have commented on earlier drafts of this paper and provided valuable guidance in its development. Particular thanks are due to Ajit Banerjee, Ram Guha, Cheryl Cort, and Prabir Guhatakurta. An earlier version of this paper was presented at a conference on South Asia's Changing Environment, which was held at the Bellagio Conference Center, Italy, from 16 to 20 March 1992.

1. Swapan Dasgupta, 'Adivasi Politics in Midnapur, *c.* 1760–1924', in Ranajit Guha (ed.), *Subaltern Studies IV* (Delhi: Oxford University Press, 1985), p. 102.
2. S. Mukerji, 'A Chapter on the Tribal Sources of Bengal History in the Muslim Period', *Bulletin of the Cultural Research Institute* 11(1 & 2): 20–4, 1975, as cited in Edward Duyker, *Tribal Guerrillas: The Santhals of West Bengal and the Naxalite Movement* (Delhi: Oxford University Press, 1987), pp. 27–8.
3. Edward Duyker, *Tribal Guerrillas: The Santhals of West Bengal and the Naxalite Movement* (Delhi: Oxford University Press, 1987), p. 28.
4. L.S.S. O'Malley, *Bengal District Gazetteer: Bankura* (Calcutta, The Bengal Secretariat Book Depot, 1911), p. 124; W.W. Hunter, *Statistical Account of Bengal*, vol. 3 (London: Trubner, 1876), p. 18.
5. W. Hamilton, *East India Gazetteer*, vol. 2, 2nd ed. (London, 1828), p. 229.
6. Binod Das, *Civil Rebellion in the Frontier Bengal* (Calcutta: Punthi Pustak, 1973), p. 45.
7. J.C. Price, '*Notes on History of Midnapore*', vol. 1, 1876, p. 1.
8. B.H. Baden-Powell, *Land Systems of British India*, vol. 1 (Oxford: Clarendon Press, 1892), p. 393.
9. See Edward Duyker's *Tribal Guerrillas*, p. 169.
10. W.G. Archer, *The Hill of Flutes, Love, and Poetry in Tribal India: A Portrait of the Santhals* (London, 1974), p. 237; as cited in Edward Duyker's *Tribal Guerrillas*, p. 169.
11. See Edward Duyker, p. 28.
12. Ibid.
13. Cited in A.K. Sur, *History and Culture of Bengal*, 1963, pp. 176–7.

14. Report from the residents of Midnapore dated 6 February 1773, as cited in Gouripada Chatterjee, *Midnapore: The Forerunner of India's Freedom Struggle* (Delhi: Mittal Publications, 1986), p. 38.
15. See note 4, p. 44.
16. Gouripada Chatterjee, *Midnapore: The Forerunner of India's Freedom Struggle* (Delhi: Mittal Publications, 1986), p. 43.
17. The British adopted the Bengali term *chaur*, meaning an outlandish or wild person, to refer to the tribal and low caste people of the area.
18. See Gouripada Chatterjee, pp. 17–18.
19. Ibid., p. 72.
20. M.C. McAlpin, 'Report on the Condition of the Santhals in the Districts of Birbhum, Bankura, Midnapore, and North Balasore' (Calcutta, 1909), pp. 20–1, 31–3, 38–9; as cited in Dasgupta, 'Adivasi Politics in Midnapore' in Ranajit Guha (ed.), *Subaltern Studies IV*, 109.
21. See Baden-Powell, *Land Systems of British India*, p. 407.
22. Midnapore District Census Report, 1951, p. 65.
23. Kailash C. Malhotra and Debal Deb, 'History of Deforestation and Regeneration/Plantation in Midnapore District of West Bengal, India' (paper presented at the IUFRO International Conference on 'History of Small-Scale Private Forestry', Freiburg, Germany, 2–5 September 1991), p. 7.
24. See Dasgupta 'Adivasi Politics in Midnapore', p. 113.
25. Ibid., p. 114.
26. F.B. Bradely-Brit, *Chota Nagpore: A Little-Known Province of the Empire* (London: John Murry, 1910), p. 4.
27. K.K. Datta, *The Santhal Insurrection of 1855–1857* (Calcutta, 1940), p. 26; cited in Edward Duyker's *Tribal Guerrillas*, p. 34.
28. See Edward Duyker, p. 35.
29. M.C. McAlpin, *Report on the Condition of the Santhals of Birbhum, Bankura, Midnapore, and North Balasore* (Calcutta, 1909), p. 34.
30. K.S. Chattopdhyaya, *Report on the Santhals of Bengal* (Calcutta, 1947), p. 35; as cited in Edward Duyker's *Tribal Guerrillas*, p. 44.
31. W.S. Sherwill, *Geographical and Statistical Report of the District of Beerbhoom* (Calcutta, 1855), p. 31; cited in Edward Duyker's *Tribal Guerrillas*, p. 146.
32. Peter Reeves, 'Inland Waters and Freshwater Fisheries: Some Issues of Control, Access, and Conservation in Colonial India' (paper presented at the conference on South Asia's Changing Environment, Bellagio, Italy, 16–20 March 1992), p. 15.
33. Ibid., p. 15, citing Sumit Sarkar, 'The Conditions and Nature of Subaltern Militancy: Bengal from Swadeshi to Non-Cooperation, c. 1905–22', in Ranajit Guha (ed.), *Subaltern Studies III* (Delhi: Oxford University Press, 1984), p. 303.
34. See Edward Duyker, p. 67.
35. 'What Happened at Naxalbari and Why?' *The Hindu*, 12 June 1967, as cited in Edward Duyker's *Tribal Guerrillas*, p. 70.

36. Interview with Pritish 'Megnath' Dasgupta, Jhargram, 10 November 1979, reported in *Tribal Guerrillas*, p. 147.
37. See Edward Duyker.
38. See Kailash C. Malhotra and Debal Deb.
39. Udayarn Bannerjee, 'Participatory Forest Management in West Bengal', in K.C. Malhotra and Mark Poffenberger (eds), *Forest Regeneration Through Community Participation* (West Bengal Forest Department, 1989), p. 4.

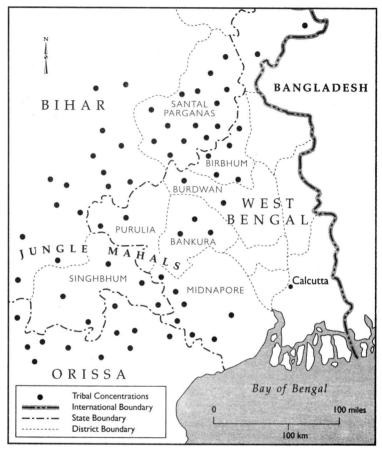

FIGURE 5.1   Tribal concentrations in the
Jungle Mahals of eastern India

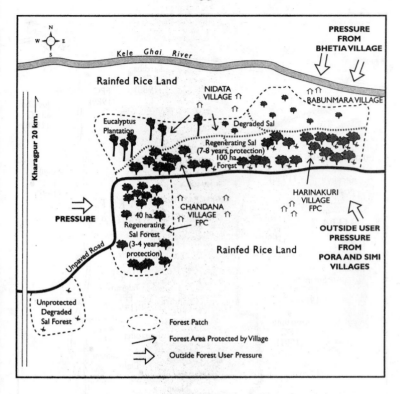

FIGURE 5.2   Forest cover and peasant activism in
Midnapore District, West Bengal: 1750–2000

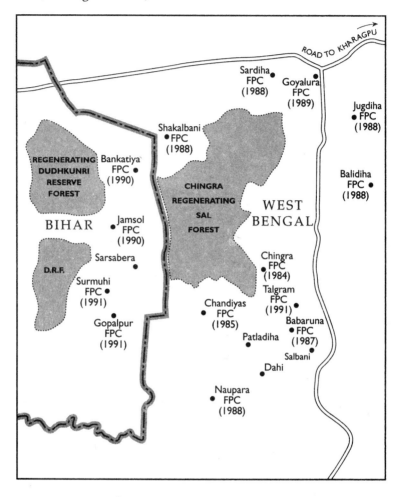

FIGURE 5.3 Chingra forest and neighbouring communities

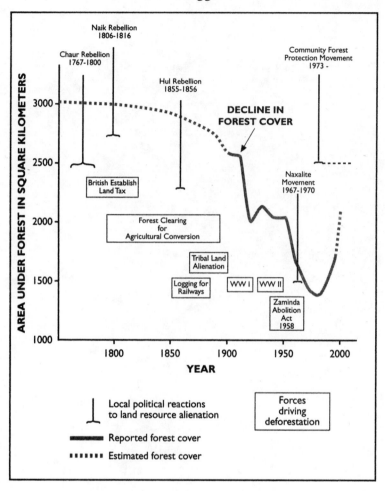

FIGURE 5.4 Chandana forest and neighbouring communities

Part II

People Transforming Forest
Management Systems

Chapter 6

From Conflict to Collaboration: Institutional Issues in Community Management

Madhu Sarin

Introduction

In many parts of rural India, communities are organizing into formal and informal groups to protect small tracts of public forest land. Sharing a common objective, these community user-groups are attempting to ensure the future sustainability of natural forests that support their manifold socio-economic needs. Increasingly, such community management activities are a response to the forest degradation and scarcities that have resulted from competing use-pressures and the breakdown of older, traditional systems of control and management. In the tribal belt of eastern India, particularly, some of these groups have emerged organically and have functioned for generations, whereas others are only a few years old. The community organizations vary in size from a settlement cluster of less than a dozen households to numerous villages comprising thousands of families. Forest management organizations also differ in structure and function. In certain Indian states, forest departments, other local government agencies, and nongovernment organizations (NGOs) have fostered the development of these diverse groups. But the key to their effectiveness is always rooted in the aspirations and capacity of the community.

Rather than offering a single, prescriptive approach to the development of community institutions, this chapter examines the nature and diversity of some of these local groups, how they define their membership and the forests they manage, the importance of basic founding principles, and the problems and opportunities they face in developing their structure and functional capacity. The essay attempts to clarify the institutional issues confronting community management institutions in order to help forest departments, local government, and NGOs better understand the types of support-roles they can play in assisting these motivated but commonly disempowered groups.

The introduction of 'scientific' forest management in India in the nineteenth century initiated the imposition of a uniform, centralized and bureaucratic management system upon a diverse range of ecosystems and indigenous management systems. Traditional communal resource-management systems existed historically in many forms throughout India. Some examples of indigenous resource-control mechanisms include the 'Kans' of Uttara Kannada, the sacred groves in the Himalayas, the 'Orans' in Rajasthan, the 'Shamilat' forests in the Punjab, the supply and safety forests in Mizoram, and the 'Cumindad' lands in Goa.[1] The process and consequences of state acquisition of India's forests have been less than smooth. Indian forest history is replete with rebellions and uprisings by forest-dependent communities against the state's attempt to deprive them of their access and control over local resources.[2]

The nationalization of forests in the post-Independence era and the forest policy of 1952 continued the process of expanding state control over forests while further curtailing community rights and authority. This was accompanied by yet another state institutional intervention, the superimposition of a national system of local governance, which established 'village' panchayats often comprising a number of socially and culturally unrelated communities. These new institutions acted as a large, single administrative unit, typically covering anywhere from one to twenty-two villages depending on the state. Whereas the all-India average sized panchayat comprises 2300 people, field experience indicates that effective community forest management groups are commonly comprised of fifty or fewer families, less than one-tenth this size.[3]

The political redefinition and the empowerment of new local governance bodies frequently undermined traditional community institutions. The state-controlled elections to *Gram* Panchayats supervised elected representatives and overlooked their financial and resource management powers. The former legitimacy of local leadership and traditions of collective decision-making were abolished and replaced by state-regulated 'representative democracy'.

These radical institutional upheavals have resulted in the progressive alienation of villagers from their older, familiar, and locally responsive management institutions. The usurping of authority by bureaucratic agencies and new, multivillage governments drew authority away from smaller, socially-cohesive groups who were closer to the forest resource, more dependent upon its stability, and better able to arbitrate access. The consequences have contributed to the vast degradation of the nation's forests.

While conflict continues to characterize the forest department–community relationship, a more favourable socio-political climate for improving forest management is emerging in India today. A growing number of planners, administrators, and foresters are beginning to recognize the need for collaborative, participatory forest management between forest departments and local villagers. Experiences from India's social forestry programmes of the 1970s and 1980s indicated that village panchayats were often too large, heterogenous, politicized, and far removed from the resource to serve as effective managers. The need for smaller groups of resource-dependent users to assume an active, primary role in local forest management is receiving increasing acceptance. Joint forest management (JFM) seeks to develop partnerships between community institutions and forest departments for sustainable management of degraded public forests. Joint management agreements are forged on the basis of trust, dialogue, and mutually-defined rights and responsibilities of both parties.

Principles of Community Resource Management

For more than a century, state forest departments have wielded enormous power over public resources with no concomitant commitment or accountability to forest-dependent villagers. For JFM partnerships to succeed, they must be rooted in *mutual* acceptance of clearly-defined rights, responsibilities, and accountability by both forest departments and community groups. For the forest bureaucracy, working with a large number of diverse and scattered local institutions will mean a fundamental shift from centralized, top-down planning and authority, to developing a capacity for decentralized decision-making responsive to the diversity of local needs and priorities. Prescriptive forest department working-plans based on technical and revenue considerations will need to be replaced by collaborative, microlevel management plans reflecting forest productivity realities and community needs. This requires reforms in forest department orientation, training, internal institutional structure, decision-making processes, and management priorities. If forest departments can learn to play the role of guide, facilitator, and technical adviser, they can nurture the development of strong, autonomous local institutions.

Participatory decision-making and decentralized management are unfamiliar concepts for conventional forest departments. Very few forest officers or field staff, and even few of the NGOs involved in JFM, are familiar with the basic principles upon which stable, durable, and democratic local institutions need to be founded, nor do they understand the type of nurturing and empowerment these community groups may require to succeed in the resource management tasks expected of them (see Box 6.1). These founding principles, based on appropriate social composition, organizational norms, accountability, conflict arbitration, and autonomy, are particularly crucial in areas where there are no strong, surviving traditions of community organization to build upon. In such situations, new traditions of collective resource management will need to be cultivated and tested, a process which is likely to be slow with uneven results. Unfortunately, the poor performance of externally imposed organizational structures on noncohesive groups of villagers, including gram panchayats, has eroded the credibility of 'village institutions'.

Community Institutions and Issues in Forest Protection

During the last few decades, three broad types of community institutions engaged in forest protection and management have developed in various states (see Figure 6.1). The first type has emerged out of local initiative, primarily as a response to the hardships of scarcity faced by local villagers due to the degradation of forests in their areas. Strong local leaders, youth groups, concerned outside individuals, including local forest department personnel, or local NGOs often act as catalysts for organizing villagers. These groups are primarily confined to areas where communities continue to have a strong economic dependency on forest produce — for subsistence or for cash — and where a tradition of communal resource management has formerly existed or is still surviving.

Thousands of community organizations, representative of this first type, are reportedly protecting more than 200,000 hectares of forests on both state and community lands near their villages in the eastern states of Orissa and Bihar[4] and, on a smaller scale, in parts of Rajasthan, Gujarat, Karnataka, and Punjab.[5] Interestingly, the forest departments of the concerned states have played only a minimal, if any, role in the development or operations of these community institutions which are characterized by tremendous diversity. While some are small and informal, others are large with formalized written rules and regulations. Leadership structures also vary, from a single, strong and trusted 'natural' leader to a more democratic and representative collective leadership selected on the basis of one person from each subgroup, hamlet, or neighbourhood. In Orissa, informal forest protection organizations include 'Groups of Village Elders', 'Village Forest Protection Committees', and 'Village Councils'. Officially registered 'Village Youth Clubs' belong to the formal category.[6] In larger villages of Orissa, it is common to find several smaller, homogeneous organizations based on the hamlet (sahi) unit. These socially cohesive hamlet groups appear to be very effective, particularly where they have successfully negotiated area-boundaries and access-control rules with the other sahis. Local institutions are also marked by various levels of organizational complexity. The shamilat forests in Punjab not only involve collective management by several villages but also reciprocal agreements with nomadic pastoral groups.[7]

Furthermore, there is broad diversity in the ages of autonomous village institutions. The sacred groves in the Himalayas, the Punjabi Shamilat forests, and the temple forests in the Western Ghats are more than one hundred years old. Although many local organizations have initiated forest protection on their own during the last two to three decades in the tribal regions of Orissa and Bihar, others have been managing a variety of community affairs for much longer. These encouraging grassroots initiatives and traditions need to be further studied and monitored closely, to improve understanding of the dynamics by which such locally derived organizations emerge, their strengths and weaknesses, and how best they can be supported and fortified.

In Bihar and Orissa, over the past twenty years, the number of community forest management groups has grown rapidly, with little or no government support. Once one community begins protecting a proscribed forest tract and natural regeneration begins, other communities observe the results and may also decide to delineate forest patches and bring them under protection. In Mahapada village in Orissa's Rupabalia reserve forest, Saura tribals settled the area three generations ago by clearing lowland forest patches and developing rainfed rice fields at the base of a forested hill tract. Bringing scheduled and cultivator caste families with them, Brahmin families moved into the area a decade later and gradually acquired all the farmland. Exercising their political influence, the Brahmins sold the forest logging rights to outsiders from Dhenkanal township. The richly forested hills were quickly reduced to scrub, while both low-income and tribal families suffered disproportionately from the loss of forest products. As a response to the hardship, the hamlet of Saura tribals approached the Brahmins with a proposal to divide the forest into user-group shares, stating 'you can cut down your portion, but let us manage ours.' The Saura began protecting a small, 25-hectare patch fourteen years ago. Rapid natural regeneration impressed the community, encouraging three other groups to form forest protection committees two years later. These forest tracts, each managed by a different subgroup that self-divided according to functional units of social organization within the community, are now more than 10 meters high. They support wildlife, trees, shrubs, climbers, and herb species, and generate significant flows of non-timber forest products. Two years ago, the Brahmins finally joined the effort and began protecting their own share of the forest.[8] The five forest protection groups now operating in Mahapada are illustrated in Figure 6.2 and their respective territories are delineated in Figure 6.3.

The second type of community organizations engaged in forest management are those promoted by the state forest departments. Older versions of these include the Forest Co-operative Societies in Kangra district of Himachal Pradesh, and the *Van* panchayats in the Uttar Pradesh hills, both initiated in the 1930s. Unfortunately, these earlier attempts at community forest management failed to truly empower villages with rights to important forest benefits. Revenues from timber were centrally controlled and channelled to the state revenue agency. Furthermore, Van panchayats were often insufficiently decentralized, comprising numerous villages that diluted accountability to the communal forest resource. The more recent goal has been to solicit villagers' co-operation in rehabilitating degraded state forests under the forest departments' more formal JFM programmes. West Bengal currently supports the largest of these programmes, with more than 2300 Forest Protection Committees (FPCs) registered by the forest department, which now protect about 320,000 hectares of regenerating public forest land.[9] Similar efforts in a number of other states, including Gujarat, Rajasthan, Haryana, Jammu and Kashmir, and Madhya Pradesh, are under way. In contrast to the self-initiated, autonomous organizations described as Type 1, the majority of the forest-department-promoted groups have been created and regulated to some degree by rules and directives specified by the concerned forest departments. The danger of this approach is the assumption that people's participation, creation of village institutions, and development of locally-relevant management options can be achieved through executive fiat. In West Bengal, where the forest department has successfully facilitated the formation of thousands of community FPCs, the programme appears to succeed largely due to their strongly-felt need and willing commitment by the communities themselves. It is doubtful whether forest departments can be effective in stimulating community organizations to adopt forest protection measures if villagers do not perceive it as their own priority. Forest departments that have played a facilitating and supportive role rather than the conventional prescriptive and directive role in forming community user groups, have experienced the greatest success in helping establish effective decentralized management systems.

The third type of community institution currently involved in forest management is sponsored by local government or NGO programmes to carry out a more general rural development mandate. In some cases, bodies such as village (gram) panchayats, women's organizations (mahila mandals), and tree-growers co-operatives have taken the responsibility for forest protection. The effectiveness of such organizations in carrying out forest management activities varies widely depending upon community interest, institutional leadership, and functional capacity. In all three types of community organizations, a commitment from a large majority of the community, as reflected in a willingness to formulate and follow access rules and contribute voluntary time to patrolling and management tasks, appears to be strongly correlated with success.

Assessing New and Old Community Institutions

Most development projects assume that they will either create a special organization to carry out project objectives or piggyback onto a pre-existing, government-sponsored village organization. Less frequently do projects take time to examine indigenous, informal institutions and organizational forms within the community. In eastern India particularly, where many spontaneous community forest initiatives have been based on clustered hamlet groups of twenty to fifty households — typically from a single tribe or clan — forest departments will need to channel support to these more decentralized traditional groups, many of whom have earned legitimacy among local residents. In such cases, special care must be taken to avoid eroding viable local institutions by superimposing new, redundant ones. This occurred during the Orissa government's 1988 crash programme when 6000 village FPCs were formed in a few months. In many cases, areas already being protected by local groups were allotted to other, more distant groups, while many of the newly created forest-department-sponsored groups proved to be ineffective. By first developing the capacity to identify traditional groups, which may function well as forest managers if empowered, government agencies can better direct their support and technical assistance services.

In cases where forest departments wish to encourage joint forest management strategies but where communities have not yet begun to organize around forest protection, government or non-governmental organizers can help assess the capacity of existing local organizations to play the role. For example, villages often have other formally recognized and/or registered groups such as dairy, marketing, savings, forest labour, or other co-operative societies. The same village may also have one or more 'committees' to assist in implementing forest or other government programmes. Where no active local organization exists, it will be necessary to support the development of a new forest management group. In either case, to facilitate evaluation, selection, or development of an appropriate community organization, it is necessary to understand the unique characteristics of JFM and the generic institutional issues that will arise at the community level, the most important of which are addressed below.

Linking People with Forests

Where rural pressures on Indian forests are high, the survival of forest ecosystems will depend on intensive access controls. Regulated entry and use can best be achieved in high dependency areas by linking primary users with specific forest territories. For communities to establish effective access controls, membership in the management group needs to be well defined and the protected forest tract clearly identified and demarcated. The process of establishing clear controls can also be a process of partial or total exclusion. Communities, NGOs, and forest departments attempting to establish more effective decentralized management systems will need to consider dependency, scarcity, residency, rights, and traditions in evolving equitable agreements among different forest-related users and communities.

Where local user groups have already started protecting forest tracts, territorial demarcations and access regulations may already be clarified among the members and with other users. In such cases, forest departments and local governments primarily need to acknowledge the legitimacy of these agreements. However, in other situations, more powerful groups may attempt to assert exclusive control over a forest area depriving other users of access, resulting in intervillage antagonism and an unstable management system. The forest department or a local NGO may need to intervene to facilitate renegotiation of the area and access stipulations between the different groups.

When outside groups or agencies assist communities in defining both the resource to be managed and the responsible management institution, the dialogue needs to be driven primarily by the users. The staff of the West Bengal Forest Department report that the dual strategy of admitting the limitations of its agency and making a frank plea to communities to take the initiative has encouraged villagers to assume the lead in management planning. Discussions seem most effective when they are open and broadly participatory to encompass the complex composite of existing forest-people relationships. Forest use patterns have been shaped by historical and cultural factors, 'rights' or 'concessions' bestowed under revenue and forest settlements, reciprocal agreements between settled cultivators and nomadic pastoralists, *de facto* usage by displaced persons or recent settlers, and customary usage by residents of distant villages. Where multiple users exist, their dependencies often vary by extent, type, and season. Unless the interests of all user groups are protected through a process of intergroup negotiation and consensus concerning the most appropriate group for taking on joint management responsibilities, the arrangement is unlikely to be stable. It may even generate new conflicts instead of promoting an improved and sustainable forest management system.

Dependency, Scarcity, and Residency

High forest dependency may be the single, most important factor in determining a user group's motivation to organize, protect, and manage the forest. Primary forest dependents may include artisans whose main raw materials such as bamboo, fibre grasses, or sal leaves are derived solely from the forest, or tribal and landless families who depend heavily on multiple wood and nonwood forest products for their family livelihood. Such an incentive is especially significant if a primary subsistence or economic forest dependence is combined with local perceptions of scarcity or threat to the resource. In the West Bengal village of Chandana, the Naik tribal community united to control access through forest protection when the villagers realized that temperatures had grown hotter, the soil was drying up, and even twigs and leaves were becoming increasingly scarce as adjacent villagers continued to overexploit the forest.[10] Whereas village residents situated further from the forest may not claim any formal rights, they may still depend on the forest periodically for collecting firewood, fodder, or forest foods. Villages closer to the forest with primary dependencies commonly seem to accept this secondary dependency, permitting access by these part-time users without generating conflict.

In addition to dependency, residence in proximity to a forest area is often a prerequisite for the practical involvement and commitment necessary for sustainable management. Communities that live close to the resource can most effectively monitor access and usually possess greater and more varied dependencies on the forest. In Haryana's pilot Hill Resource Management Societies (HRMS) of Sukhomajri and Nada, formed in the early 1980s, communities decided that permanent residence in the villages next to the forest was a non-negotiable condition for HRMS membership. Non-resident landlords were appropriately disqualified from membership since they could not actively participate in improved forest protection. This also ensured that nonforest-dependent individuals living in urban or distant settlements could not usurp control over forest resources on which local residents were highly dependent.

Rights and Tradition

As they organize, community groups and assisting agencies must consider formal and informal resource rights. Most Indian communities carry with them a long legacy of resources rights and traditions, some of which date back centuries. Many of these rights were recorded by the government under forest settlement acts that were negotiated in the late nineteenth or early twentieth centuries. Other agreements were informal acknowledgments of prior rights or settlements made between neighbouring villages to resolve or prevent resource conflicts. In most cases, poor documentation and differing perceptions of rights exist. As a result, the process of working out mutually acceptable agreements to establish community access control systems and management mechanisms often encounters complex questions of prior rights and traditions.

When the Haryana Forest Department began accelerating its joint forest management programme, certain residents of adjoining and distant villages protested that joint management agreements were being exclusively negotiated with only one of numerous traditional user groups. In some cases, those more distant claimed equal or greater traditional rights to the forest area than the closer villages. Some users claimed their rights based on earlier residence in the area, while others based their claims on prior agreements with the forest department or other government agency. Several distant communities asserted product-specific rights (e.g. to fodder) based on past usufructs.

Several agreements drafted with single villages in Haryana had to be modified later to eliminate conflict produced by the unintended exclusion of other user groups. In Lohgarh village, the HRMS entered into a lease agreement with the Haryana Forest Department for the joint management of fodder grass compartment C3 in 1983. Income from the annual leases went into the HRMS account to support general village development. Eight years later, leaders of the adjoining village of Manakpur Thakur Das (MTD) protested their exclusion, claiming that their village enjoyed exclusive rights over one portion of C3. While the colonial period forest settlement agreement of 1938 mentions no such demarcation of territory between individual villages, a joint meeting of the two villages resulted in Lohgarh residents accepting MTD's claim. The boundary pillars mentioned by the MTD villagers were also found in the field, further validating their claim. Probably granted during the area's earlier revenue settlement as far back as 1908, accommodation to these rights proved essential. Haryana's JFM support team facilitated renegotiation of the boundaries and use agreement for compartment C3 by organizing joint meetings of the two villages. Based on consensus, MTD formed a separate HRMS and began leasing its share of C3 in 1991 (See Figure 6.4). Subsequent problems related to lease pricing were similarly resolved through open, consultative discussions between the two villages.

In other Indian states, villages located far from forest areas have legal settlement and nistar rights. While it is difficult for them to participate in day-to-day protection and management, they cannot simply be excluded from a JFM agreement unless either the rights concept is changed or distant right-holders agree to delegate forest management responsibility to more proximate groups on mutual terms of access. The inability to negotiate mutually agreed demarcation of boundaries can result in the collapse of the local organization as well as destruction of the regenerated forest. In some cases, with self-initiated forest protection groups in Orissa and Bihar, and even the formally recognized FPCs in West Bengal, communities alienated from the resource over which they perceive themselves to hold rights have conducted mass loots.[11]

In western and northern India, conflicts between resident users and nomadic pastoral communities can arise. Pastoral nomads include breeders and herders of camels, donkeys, yaks, pigs, sheep, goats, buffaloes, and cows. More than 200 castes numbering up to 6 per cent of the total Indian population are engaged in pastoral nomadism.[12] Practised for thousands of years, migratory pastoral groups, typically, have clearly delineated routes and usu-fruct agreements with agricultural communities, many of which have endured for generations. However, pastoral communities have suffered greatly from the progressive decline in the country's common lands. The shrinkage and degradation of common prop-erty resources have forced large numbers of nomads to give up their traditional livelihood and join the ranks of the urban and rural poor. Professional foresters view these nomadic graziers as a major threat to the forests. More than a century of legislative acts have attempted to curtail or deny their grazing rights on forest lands, yet, in many regions these pastoral communities continue to travel with strong seasonal dependencies on forest lands — whether *de facto* or *de jure*. These realities need to be carefully considered by forest departments in their efforts to empower local user groups as forest managers.

In the Shiwalik hills of Haryana, traditional right-holders have been unaccommodating. While the prosperous Jats of Jholuwal claim traditional rights in the adjoining forest compartment, they depend on it only marginally to supplement fodder collected from their private holdings. Residing closer to the forest, a more recently settled community of Banjaras depends heavily on forest *bhabbar* grass for rope-making (*baan*), their primary occupation. Despite a series of meetings with forest department community organizers, the Jats refused to allow the Banjaras to buy the bhabbar grass lease. As no consensus could be reached, a joint management agreement was stymied.

In an effort to minimize future intergroup conflicts in the joint management of a forest area, the importance of equity surrounding tradition must be weighed. By accepting traditional use patterns as the primary determinant for the selection and extent of forest area allocated to certain groups, reliance on his-torical accident — rather than need or management capacity — may become the basis for allocating usufruct rights to a common property resource. Although traditional right-holders may indeed be the logical managers and legitimately most in need of the resource, in other cases, small but politically more powerful groups may demand exclusive management rights to large forest tracts simply on the premise of having enjoyed rights there earlier. This monopoly may be at the cost of the much larger and poorer adjacent populations.

Apart from traditional users or right-holders, a large number of the poor in India are displaced each year due to large dam and related infrastructure projects, famines, and other natural disasters. Settling into new areas, they commonly become heavily dependent on forest resources. In some situations, the right-holding community is reluctant to extend usufructs to migrants, despite their obvious and often greater resource needs. As forest departments enter into joint management agreements, they may face a dilemma in helping determine whether higher priority should be given to older, traditional right-holding communities or to migrants with stronger forest dependencies. In Limbi forest in southern Gujarat, the original tribal settlement expanded rapidly in the 1970s when migrants from neighbouring forests arrived after their land had been inundated by a new dam project. Rendered landless, with no employment alternatives, the majority of displaced women have now become highly dependent on daily commercial fuelwood headloading from the Limbi forest. While the forest department and local residents are allowing the women to carry out substantial fuelwood loads, overexploitation steadily degrades the forest. To avoid future conflicts, joint management agreements in the area will need to accommodate such 'ecological or development refugees', at the same time taking into account the rights of older families.

To better identify forest user groups, the Haryana Forest Department has found that simple mapping techniques serve as an effective planning tool. Existing use patterns are mapped with members of different groups to illuminate formal and informal rights and resource dependencies. This information provides a sound basis for discussion regarding the terms of the management agreement. Increasingly in Haryana, user groups living at distances of three to five kilometres from the forest, and therefore unable to participate in daily protection and management, have been allotted secondary usufructs. Primary users, who tend to reside nearer to the forest area, enter into a joint management agreement with the forest department and accept primary management responsibility, while permitting continued access to the non-member secondary users on clearly defined terms. This compromise strategy helps prevent the inequitable exclusion of periodic benefits to the more distant secondary users, while also formalizing access rules and clarifying the range of user rights in the villagers' minds.

Ultimately, a degraded forest ecosystem can only regenerate if it is protected, with limited extractive pressures, for a certain period of time. Increasing pressures from new user groups pose additional ecological constraints. With India's growing displacement and population migrations spurred by developmental projects, environmental deterioration, or new economic opportunities, the challenge is to strike an intricate balance between the regeneration of degraded ecosystems, the preservation of traditional rights, and the redistribution of forest access on the basis of greater equity. This will require a much better understanding of sustainable forest-use levels, historical and socio-political dynamics, and forest dependencies of the full range of past, current, and potential user communities. If consensus can be achieved, local user groups can unite in a community management organization that has a far better chance of sustaining the forest than community and government factions that misunderstand each other and continue to compete for a dwindling pool of finite resources.

Evolving Structures in Community Institutions

Experience indicates that the most effective local resource management groups operate democratically, with relative self-sufficiency and independence at the grassroots level. Universal membership is common; however, the main body may be governed by a managing committee of representative leaders. It is important to highlight the difference between the ordinary members, constituting the main body of the organization, and its representatives or leadership. A 'committee' comprised of so-called 'representatives' cannot in itself comprise a local organization without a general body of members whose interests it represents. Yet a number of state-sponsored joint forest management (JFM) programmes refer to the term forest protection 'committee' with no reference to a general membership, thereby equating a committee of a few individuals with a community institution. To be democratic, management committees must represent the will and priorities of the larger set of user groups. In turn, these groups must reach a consensus, maintain a clear identity of membership, and fully understand their rights, responsibilities, and rules.

General Membership in the Management Institution

The principal function of a community institution in JFM is to provide an organizational structure that can articulate and represent the interests of *all* user subgroups of a forest area in partnership with the forest department. This can only happen if each group's interest is adequately represented in the institution. The community institution should also be able to facilitate intergroup negotiation and consensus to balance the relative costs and benefits of various forest management options. Ideally, all general body members should feel that the community institution will safeguard their legitimate respective interests.

Such partnerships have proven easier to negotiate with community institutions consisting of homogenous groups where members share a similar socio-economic status and dependency level on the forest. However, heterogenous groups, not only in social and caste terms but also in the degree and nature of forest dependence, are more typical. They often face more difficulty in agreeing and organizing. A joint management agreement acceptable to one user subgroup may be against the interests of another, or at least be perceived as such. For example, an agricultural subgroup may be interested in enforcing a total grazing ban in the forests to protect its irrigation tank from siltation. Another user subgroup of local or nomadic graziers sharing the same forest area may find this totally unacceptable, unless provided with a viable, alternative grazing option. The immediate opportunity costs of a grazing ban would be very different for the two subgroups. Unless a mutual agreement can be reached on how to compensate fairly the higher costs borne by one, resource stability through JFM is likely to remain elusive.

Traditional power relations and perceptions about the relative superiority of different groups in the social hierarchy also need to be considered while studying or designing community institutions. Very often, the subgroup most dependent on forests is also in the lowest socio-economic bracket. The community institution must be able to ensure equal representation of the interests of the most disadvantaged minorities in negotiating a partnership with the forest department. Otherwise, if left to the identified user group as a whole, the most dominant subgroup within it may easily appropriate control of the community institution to serve its own interests.

Thus, the constitution of a representative and stable community institution capable of performing its forest management tasks needs to be facilitated by an *empowering* and *participatory process.* This evolving process must include open discussion and negotiation among all those likely to be affected by the JFM agreement. Certain criteria can be useful guidelines in determining eligibility to the community institution's general body membership. These include voluntary, open membership to all resident users; ensuring equal and independent eligibility for women; option for future inclusion of new settlers; and an ability to attend meetings at short notice.

Representation and Eligibility: In principle, eligibility to membership of a local organization should be open to all right-holders/ users of the concerned forest tract, irrespective of the extent and nature of their dependency on the forest. Restricting eligibility to individuals or households selected on the basis of caste, tribe, class, or economic status tends to prove divisive, often creating resentment and conflict. Selective membership criteria were initially adopted by the 1989 West Bengal resolution. In 1991, due to objections raised by those excluded, membership was opened to all households 'resident' in the vicinity of the forest area, subject to their being interested in forest protection. In states where government programmes require that membership be open to all, it is usually based on residence in the vicinity of the forest area. However, the definitions of 'residence' and 'vicinity' tend to be vague and arbitrary. For example, the Jammu and Kashmir resolution specifies residence 'at the edge of the degraded forest' as a membership qualification, whereas the Rajasthan resolution offers membership to all those resident in the 'revenue village' adjacent to the allotted land. Residents within a revenue village may only partially overlap with a larger sphere of user groups.

Other state orders mention a village, a group of villages, or the panchayat as the qualifying place of residence. Although the West Bengal amendments are a positive example of response to learning from the field, such externally imposed membership prescriptions hinder, rather than promote, organic development of effective community institutions.

While some users have only a marginal interest in forest produce, a particular subgroup may be totally dependent, such as Gujarat's Kotwalia bamboo basket-makers who depend on access to regular and ample supplies of raw materials. While those with marginal dependencies will have little motivation for investing additional time and effort in improved forest management, others have a high stake in maintaining and increasing productivity. Marginal users may choose not to become members of the organization, provided their forest access for meeting limited needs is unhindered. Such a decision should ideally emerge from the larger group, with those opting to decline membership giving their consent to the more dependent users to constitute the community institution.

The criteria of universal eligibility in an agreement or resolution is often inadequate in itself, particularly in groups with high levels of socioeconomic disparity. It is most important to ensure that all those eligible are aware of their right to join the organization, and that the most underprivileged, especially women, are encouraged and empowered to do so. Careful monitoring can ensure that no household or subgroup is being *wilfully* prevented from membership. As the larger and more powerful of the two partners, the forest department must be firm about demanding compliance with such generic principles. This will facilitate the breakdown of traditional barriers that inhibit closer co-operation and interaction between different subgroups, particularly the more disadvantaged users.

Women's Role: Outside agencies can contribute significantly by promoting women's independent eligibility for membership in emerging community resource management institutions. When the issue of women's participation in JFM is placed on the agenda, it automatically provides opportunities for them to play a role in institutional decision-making. Where the formation of new groups is being promoted by forest departments, they are in the advantageous position of introducing progressive changes in India's traditional organizational forms by insisting that priority be given to equity and gender issues in the new institutional structure and functions. Gender equity from the outset is often avoided or cautioned against on grounds of resistance from conservative, tradition-bound village men. However, if the community has a high stake in a benefit-sharing partnership with the forest department, there tends to be surprisingly little resistance to the acceptance of gender equality as a founding principle of a new community institution. This has been witnessed in Haryana, where women's traditional socio-economic status is among the lowest in the country. Irrespective of religion or caste, few village men have questioned the forest department's insistence on treating women as equal and independent members in HRMS as a precondition to finalizing a joint partnership agreement. In states like Orissa and Bihar, where autonomous forest protection groups have been functioning for many years, attempts' of the forest department to encourage gender equality have been less successful and may require a more concerted effort.

Despite the diversity of Indian cultural traditions, women rarely play a role in decisions related to community affairs. There are few, if any, traditional forums that enable women to get together to discuss and share their problems. Not surprisingly, most community institutions that have initiated forest protection on their own are comprised of only male members. A case study of two self-initiated forest protection groups in Orissa revealed that women were not *allowed* to attend general body meetings, even when they were involved in the dispute under discussion.[13] Yet, according to the traditional gender-based division of roles in the majority of forest-dependent communities in India, the women are the main collectors, users, and processors of non-timber forest products. Apart from considerations of gender-equity, this factor alone demands women's participation in JFM.

Unfortunately, the majority of government JFM resolutions do not address this aspect of general body membership. By prescribing eligibility for one representative per household, most of the orders automatically exclude women — except for all-women households or widows without adult sons. Using the typical family or household as the qualifying unit for membership, fails to deal with the diverse variety of situations pertinent to a growing number of rural women. According to the 1981 census, 8.5 per cent of adult women in the country were either widowed or divorced. Research has also shown that in 26 per cent of the families surveyed, women were the sole source of household income.[14] With the obvious rapid pace of socio-economic change in the country, an increasing percentage of women are effectively becoming the primary supporters of their dependents. Such households are disproportionately concentrated among the population living below the poverty line. Factors such as male-dominated migration, unofficial abandonment of women, men taking second wives to beget sons, male unemployment or disability, or women choosing to remain single, are resulting in the growth of such matriarchal households. Since many such women form invisible subunits within larger households, it is important to entitle them to an independent share of usufruct and income benefits through JFM. Reaching out to these women can be facilitated by opening community institution membership to all adults, female and male.

Migrants: With industrial and other economic activities expanding near forest areas, inevitable in-migration of new settlers searching for employment will continue. While the more advantaged newcomers depend less on forests, factory or casual wage workers tend to supplement their low incomes by collecting subsistence goods such as firewood and fodder from the forests and common property areas. Although dependent on the forest, their commitment to investing time and effort in improved forest management may be limited due to their temporary residence. In such situations, whether they should be eligible for community institution membership or not becomes an issue. Instead of prescribing to any one rule, traditional residents should be encouraged to reach an appropriate case-by-case decision through discussion and consensus.

The issue of new settlers as members of the HRMS arose in relation to the eligibility of employees of a cement factory located near Pinjore, Haryana. The residents of three different villages (Surajpur, Rajjipur, and Manakpur Nanak Chand) share certain rights in the adjoining reserve forest. The families of some factory workers also collect fodder grass from the forest for their cows or buffaloes kept for domestic milk consumption. To start with, a small group of enterprising factory employees from two of these villages formed a registered society and successfully lobbied to obtain a fodder grass lease from the Haryana Forest Department (HFD) for two forest compartments.

During follow-up meetings with the society representatives, Haryana Forest Department's JFM support team encouraged the small group to increase and open their society's membership. However, Haryana's principle of universal eligibility of all resident adults proved totally impractical in this case. Although the combined number of households in the three villages does not exceed one hundred, the factory settlement has more than 1000 families. Encouraging all adults to join the society would have made it an unwieldy and unmanageable body, with disproportionately low representation of older residents compared to the factory workers. Discussion of the issue in a society meeting led to a simple solution. The members agreed that irrespective of gender, period of residence, or occupation, only those persons who actually go to the forest to collect fodder would be entitled to membership. This condition effectively screened out those with no direct dependence on the forest, thereby making the HRMS an organization of actual forest users.

Management Committees

There is some debate over the need and value of a separate management committee, which wields greater decision-making authority than the general body membership. Traditional community organizations have evolved over a long period of time, and many have developed responsibilities and codes of conduct for their leaders. In some cases, the leaders' primary role is to function as facilitators in helping the community reach a consensus regarding management decisions. While these roles and codes of conduct may not have been formalized, they often effectively serve to guide the leaders' actions. Certain community organizations, particularly the older, smaller, and traditional types that tend to operate with high levels of accountability, often reach decisions based on consensus by the entire membership body, or representatives of various subgroups comprising an informal committee or council of elders. In other small and homogeneous traditional communities, the task of daily operational decision-making may be performed by a single, strong leader whose judgment is trusted by others.

In newer community resource management institutions, when membership expands beyond a certain size, there is often a need to delegate powers of day-to-day decision-making to a more formalized structure of a management or executive committee in which a number of elected or selected office bearers are supported by nonoffice-bearing representatives. Interestingly, many of the self-initiated community institutions in Bihar and Orissa have evolved their own rules for constituting such management committees.[15] Where no indigenous traditions of communal management exist, or where earlier traditions have been weakened, the roles and responsibilities of both general body members and their management committee need to be clearly defined by all parties. Unfortunately, this has commonly been ignored in the development of new local institutions promoted by forest departments for JFM. Although formal structures are designed and individuals appointed to fill various positions, neither the general members nor the managerial representatives are informed properly or empowered to play their roles as effective functionaries. This is particularly true where management committee members are officially imposed upon a community from outside without any consultation process with those whose interests they are supposed to represent. If the leadership lacks accountability to the community it represents, a community institution tends to be ineffective in performing its tasks, resulting in a serious loss of credibility.

Governing Principles: A basic set of governing principles can help ensure the constitution of an equitable and effective management committee (MC). Only the general body of the community institution should be empowered to elect the MC. In lieu of state resolutions that prescribe static, numeric representation for gender and socio-economic groups, it will be more effective to follow the principle of equal representation of all parties. The forest department or the local NGO can help monitor the selection process to ensure that the weakest groups are not sidelined by the more dominant ones. Special efforts will be required to promote women's representation on the MC in view of their traditional exclusion from such roles. A non-voting outsider, such as a respected school teacher or village elder, may be inducted as an ex-officio member to play an advisory or conflict-resolution role. A forest department representative can serve as a liaison between the agency and the village, keeping the group informed about forest department policies and programmes. Selection of an outside MC representative by village members is best guided by such factors as the degree of community trust, credibility, local availability, and commitment to the cause.

It is crucial that the process by which MC members are selected or elected is transparent. In more stratified groups, facilitation by an outsider may be required to guarantee that disadvantaged sections have a genuine opportunity to influence choices. When a particular department is promoting a community institution for one of its specific programmes there is a logical tendency for staff members to select the office bearers or, at least, to strongly influence their selection. If an agency staff is insensitive to the political dynamics between the stronger and weaker sections of a community, and assumes that representatives of the stronger group are the most suitable candidates, the weaker subgroups will lose their representation in MC decisions. Forest department field staff must be sensitized to the importance of empowering the traditionally weaker subgroups to articulate their priorities. Another danger is the concentration of MC membership among a few families. This often leads to the group splitting into antagonistic factions, and eventually collapsing. To avoid family-monopolized membership, the general body may wish to restrict MC membership to no more than one member from each family.

Consensus-based 'selection' of MC members in an open general body meeting may not be fully representative since traditionally disadvantaged groups, including women, are commonly too inhibited or intimidated to participate fully. A compromise solution may be for each forest-related interest group to select its representative during smaller subgroup meetings. Such a tradition already exists in some areas, particularly among the indigenous forest protection groups in Orissa and Bihar. In Haryana, the typical pattern of MC member elections is still exclusive, male-biased and top down. A small group of dominant men move away from the general assembly, debate among themselves, and return to announce the names of their proposed MC members, including the designated office bearers. More recently, the Haryana Forest Department has included the additional stipulation that at least two women should be selected as MC members. Yet village women are seldom consulted in choosing their own representatives, unless facilitated by an outside team. Those females selected as MC members tend to be the wives, mothers, or daughters-in-law of male MC members.

At the same time, voting by secret ballot or campaigning for election is seldom practiced. Its merits are questionable as it can easily divide the community into factions, with elections becoming an issue of prestige and patronage. Such negative consequences have already occurred with panchayat elections in many Indian villages. The development of a community institution — organized around issues such as JFM, and characterized by equitable decision-making mechanisms and truly representative leadership — requires the creation of a conducive, non-threatening environment that encourages dialogue, debate, and consensus among all subgroups.

The accountability of an MC (or a less formal leadership structure) to the general body is one of the key determinants of a community institution's effectiveness and stability. Research indicates that both indigenous and new community institutions (CIs) begin to break down when the majority of members lose confidence in their MC or the informal leadership, and are unable to find any channels for redress other than to simply withdraw from participation. This is particularly true when the CI is handling forest-based income and suspicions develop about misappropriation of funds. A variety of institutional mechanisms can be adopted by the CIs to help minimize the chances of such collapse. The first is to institutionalize regular consultations with the general membership body to ensure an accountable decision-making mechanism. Too often it is falsely assumed that as the CI's 'representative', an MC member automatically understands what is in the community's best interest. In fact, many MC members may have assumed a public decision-making role for the first time. With no experience or clear notion of their responsibilities, they may remain indecisive or prone to misguided decisions. To avoid such a situation, the mandate of the MC must be defined as clearly as possible in consultation with the general membership, and its decisions must be regularly monitored. While the MC may be empowered to make certain types of decisions or commitments on behalf of the CI, it must be obliged to obtain the general body's approval before finalizing major decisions likely to directly affect the CI's membership. The latter would include matters such as approval of a joint forest management plan committing all members to halt grazing for a specified period, the terms for obtaining a grass or fodder lease from the forest department, or incurring expenses above a certain amount from the CI's common fund. Unless such important decisions are based on prior approval from the general body — permitting general members the opportunity to not only accept or reject them, but also to suggest modifications — the MC can rapidly lose the general body's confidence.

To operationalize accountability mechanisms, it is useful for the CI to formally schedule a minimum of two, preferably more, general body meetings each year. These meetings provide an opportunity for the MC to inform members about its activities during the intervening period, present an income and expenditure statement, discuss problems, and seek membership approval for future activities. By binding the management leadership to regular consultations with the general body, such meetings provide a forum for ordinary members to raise issues so that any disenchantment or resentment can be expressed early and remedied. Similarly, consultations among MC members about proposed plans and decisions are also helpful if structured regularly. In areas with a tradition of community organization, membership meetings once or twice a month are a typical practice. Still, these are generally exclusively male forums and urgently need to be adapted to accommodate women. Introducing the norm of regular meetings where no such prior tradition has existed may initially require considerable encouragement and guidance from outside facilitators.

For a genuinely representative and participatory CI, its formal executive committee or informal leadership should be able to consult the general membership on critical decisions, sometimes at very short notice. If the membership is scattered over a large area making such meetings impractical to organize, it may be more desirable to restrict membership to a smaller, more cohesive physical unit to whom others are willing to entrust the responsibility of day-to-day decision-making and management. If this is not acceptable to the other user groups, then formation of a number of smaller organizations may be considered. Determining which subgroups or physical subunits should form part of which institution is best left to the villagers to decide. User groups should determine management responsibilities and an equitable benefit-sharing mechanism, with the forest department playing a monitoring role when necessary.

Evolving Functions of a Community Resource Institution

The success of local forest protection and management activities is most often determined by the routine functional capacity of the CI. Forest management functions can vary from simple protection through patrols, to the formulation of complex micro-plans that enhance productivity — i.e. enrichment planting and silvicultural manipulations to maintain diverse product flows — for meeting community needs. To manage the forest resource effectively, it is imperative that the CI and the forest department reach a clear agreement regarding management objectives.

Defining Management Objectives

Community-led decisions to initiate forest management are often triggered by rapid or dramatic environmental changes. These may include landslides, dam bursts and flooding, droughts, illegal logging, departmental fellings, forest fires, loss of agricultural land and/or productivity through erosion, depletion of forest produce for subsistence or cash livelihoods, desiccation of water resources, or encroachment of forest areas. In most areas where community groups are taking the initiative to protect degraded natural forests, their goal is to protect forests from excessive exploitation and encroachment by outsiders, and to expedite natural forest regeneration. The community's objectives are to recover forest stability and ensure sustainability in order to maintain the steady flow of wood and non-wood forest products upon which they rely.

Due to their greater political power, forest departments can pressurize CIs to adopt their own agency objectives, either consciously or inadvertently. Many forest officers still feel that according to historical forest-settlement agreements, community 'rights' to forest products should be limited to their '*bona fide* domestic needs'. When village forest managers wish to sell surplus forest produce, forest department staff may object. In degraded areas requiring enrichment planting, forest department officers, due to their traditional orientation, may lobby for timber species, overriding community priorities for fodder or fruit-bearing trees. As a result, the subsistence needs of the less powerful households in the village can be overlooked.

In many social forestry projects implemented in recent decades on community lands, forest department and donor preferences have resulted in the planting of fast-growing, commercial tree species. While it was intended that these species would provide basic firewood and fodder needs to the villagers, the species were frequently not suited for those uses, nor did villagers — unconvinced that they would benefit in any way — participate in protection or management. One-time timber harvests were often auctioned to outsiders, and the benefits were typically usurped by panchayat elites who did not represent the village constituency. In the process, the villagers most dependent on panchayat lands for collection of subsistence goods lost access to the enclosed commons, as well as control over the final income. Under JFM programmes, both forest department staff and local NGOs will often need to play an important support role in assisting village institutions to identify the wider community's forest management objectives.

Access Controls

A primary function of a forest management CI is to control access to the resource. If this does not place any major restrictions on community member activities, consensus on the necessary rules for restraining outsiders will be relatively easy to achieve. On the other hand, if controlling access requires complete closure for a grazier subgroup or a total ban on fuelwood headloading, reaching consensus will be much more difficult if viable alternatives cannot be provided. A dominant subgroup may still succeed in enforcing closure to suit its own interests, but the system is unlikely to be sustained due to the unresolved, albeit suppressed, conflict of interests within the membership.

While many of the self-initiated forest protection groups in Orissa and Bihar are committed to enforcing access controls through openly democratic discussions, in some cases overzealous subgroups have enforced such controls through various types of pressure and without consideration of equity issues. For example, certain youth clubs in Orissa have imposed a total ban on firewood collection from community-protected forests, simultaneously appropriating the right to sell firewood collected during cleaning operations, to generate funds for a new youth club. These actions have deprived the most disadvantaged, forest-dependent members of access to an essential fuelwood source.[16] Similarly in Bihar, two of the ten self-initiated forest protection groups studied had imposed the management objectives of the more powerful minority subgroups on the majority of village members.[17]

A coalition of individuals cannot achieve a common resource management objective without a large majority of its members consenting to follow the mutually agreed rules that regulate their resource use behaviour. Villagers have used a range of approaches to shift 'open access' forests to more closely managed and regulated areas. Strategies generally require a consensus among all or the majority of community members regarding the establishment of boundary demarcations, patrols and watchers, fines, extraction fees, and seasonal limits. However, the process of evolving consensus on rules that regulate access may be highly sensitive, requiring months or even years.

Once a community group has decided to begin protecting a forest, members must agree on the type of access control system to follow. Provided there is a strong consensus for forest closure and some mechanism for compensating those incurring a disproportionate share of immediate costs, peer group pressure is often most effective in controlling the behaviour of members of the immediate community. Nevertheless, peer group pressure may need to be reinforced by specific punitive measures, including fines or other penalties, that serve as deterrents. During the initial years when closure is first imposed and natural forest recovery is in its early phases, most communities need to employ a system of paid or voluntary patrols. Communities in Gujarat and Orissa have developed a system of stick rotation known as *tengga pali*. A bamboo or any other pole is rotated from household to household each day according to which family is responsible for patrolling the forest. Patrolling may be done by a single adult male, or by teams of two or three, including older women. In general, patrols operate only during the day when grazing or fuelwood cutting is most common, although where the community feels the threat of a timber raid after dark, night watchers may also be necessary. A group of thirty-five tribal villages, which collectively protect a forest tract in southern Rajasthan from organized gangs of timber smugglers, employs one able-bodied person from each family. On hearing a special call by one of the village patrols, all households immediately rush forward to assist in turning away the raiders. All families are bound by this rule, irrespective of whether the alarm is raised during the day or night. In the southern Gujarat village of Gamtalao, community members maintain active daily patrols during the first two years of forest closure in order to discourage neighbours and nomadic Rabari herders from damaging the forest. Gradually, outside users learn about the community's commitment to protection of the forest tract and come to accept the limitations on access. After outside attempts at exploitation decline, the community shifts to a less intensive system of watchers.[18]

Such informal watchers are frequently used to monitor access. This system requires less labour than patrols, but also provides less thorough access regulation. Women who collect non-timber forest products or take animals to graze can frequently serve as *de facto* watchers. If a woman headloader enters the protected patch, a female watcher may warn her that the forest is closed to cutting. Should a larger group arrive from another village to harvest fuelwood, she may need to call the village men to turn them away. In Sukhomajri in Haryana, the concept of 'social fencing' entailed an agreement by all HRMS members to exercise voluntary restraint on forest grazing and to keep watch for any offenders without using organized patrols. Villagers adopt various ways to deal with offenders, including verbal warnings, fines, threats of social ostracism, or turning them over to the forest department. Warnings seem to be used most commonly, especially with first-time offenders. Fines are often levied on members with previous records of warnings. Imposing fines or other forms of penalties on non-members, however, can be problematic as it raises the issue of the CI's legal authority. Court cases have been filed against forest closure enforced by autonomous community groups;[19] the president of the protection committee of Hardatal in Orissa was arrested because the group had detained cows that had entered their forest.[20] Similar problems are encountered by groups when they attempt to close access to forest tracts in which other villagers also have rights. A useful approach is to build up support for closing access among these neighbouring villages. The Gamtalao FPC in Gujarat sent representatives to an adjacent village to meet with the elders of an offender. The elders agreed to extract a fine of Rs 120 ($4) from the violator, for illegally grazing his animals in Gamtalao's forest.[21]

Communities which work in active collaboration with the forest department, are more prone to turning over repeat offenders to the local field office. In Salboni village of West Bengal, the FPC had repeatedly warned an individual about cutting the forest for commercial firewood. After discovering him loading a vegetable cart with fuelwood one day, a group of fifteen FPC members dragged the offender, along with the cart, to the range office guard. Even though the guard eventually let him off with a small fine, the offender, who was humiliated, promised to reform. In Harda Forest Division in Madhya Pradesh, the village protection groups work in close collaboration with the forest department field staff to book offenders. During the five-month period from November 1991 to March 1992, the forest department collected Rs 6150 in fines in the village of Khardana.[22] Sometimes, if the villagers feel that the official fine is minimal for the gravity of the offence, the forest department staff informally permit the group to levy a higher fine. Although the official fine is deposited with the forest department, the rest is deposited in the CI's common fund.[23] Villagers seriously enforce rules that they are committed to. For example, the protection committee of Domra in Harda Division even fined the forest guard for failing to perform his protection duties.[24]

Trouble arises when forest department field staff fail to provide promised back-up support to CIs which are promoted under state JFM programmes. Since the CIs are not legally empowered to punish outsiders it may result in a serious loss of credibility for them. Although the HRMS of Harijan Nada in Haryana repeatedly brought offenders to the attention of the forest department field staff, they avoided taking action and the HRMS's moral authority was progressively eroded. If the CI is unable to regulate the access of outsiders, over time, even its own members begin ignoring the rules framed by the community. The system can thus rapidly revert to an open access situation.

In certain cases, confrontations become violent and offenders are occasionally beaten. 'Mass loots' occur where dozens and even hundreds of villagers from outside communities join *en masse* to fell a regenerating forest or plantation during the night. Often protection group members gather with poles, spears, and bows and arrows to resist them. But when local groups are outnumbered, the appearance of local forest department staff in support of the protection group can effectively break up the looting. In the face of such forceful outside pressures from neighbouring villages or powerful commercial interests, visible forest department support can be the critical factor in protecting the rights of the community protection group and ensuring the survival of the forest.

Harvesting and Benefit Distribution

Except in cases where communities are strictly protecting forests for soil and water conservation, biodiversity, or environmental amelioration, an important function of CIs is to devise rules that regulate the harvest and collection of forest produce by its members. The more economically dependent CI members are upon the forest, the more important the rules. While ideally ensuring that users' needs are met both adequately and equitably, harvesting rules must also ensure sustainable levels of exploitation. Since these exact levels are unknown, a careful process of trial and error with mid-course adjustments must be adopted. The extent and periodicity of extraction for different products must be adjusted against the changing conditions of the forest ecosystem. Contrary to the common perception among foresters that villagers are unable to practice 'scientific' or sustainable forest management, many autonomous village groups have developed sensitive and sophisticated resource extraction rules. When village members are concerned enough and highly motivated to protect their resources in perpetuity, they have a strong incentive to devise harvesting rules that ensure the recovery of biomass, which equals or exceeds its extraction.

Rules developed by CIs in Orissa and Bihar fall into two categories. The first type aims at facilitating natural regeneration and curbing excess use of valuable or scarce products. The forest protection group of Chargi village in Giridih district of Bihar prohibits timber extraction more than once every three years for certain agricultural or irrigation implements. Through social pressure, members are encouraged to properly maintain and prolong the life of their implements. Felling timber for sale or distribution to relatives from outside villages is forbidden. The CI of Buzurgnano village in Hazaribagh District has similarly banned the use of timber for fencing to minimize non-essential use. As an alternative, members are encouraged to use only Lantana, thorny bushes, or dead wood for fencing. For some types of uses, replacement of timber by bamboo, which has a faster regeneration cycle, is recommended. Other rules designed to curb wasteful consumption and facilitate regeneration of useful forest produce include a ban on cutting young saplings or lopping the main buds of young trees; a ban on felling of select tree species except with prior permission of the leadership; a ban on uprooting of the aggressive Lantana shrub, and supervised grazing in areas that permit it. Community groups in Bihar have also begun felling coupes that they have demarcated themselves. Grazing in the felled areas is banned for two to three years to facilitate natural regeneration.[25]

The second type of rules comprise time-tested, older indigenous management systems, as well as newer community initiatives, and employ strategies to ensure timely and sustainable collection, harvesting, and division of forest produce. The second category of harvesting rules regulates extraction of forest produce from community protected areas as a mechanism to distribute benefits. These involve the allocation of specific rights to individual tree or plant species, to collection areas or ranges, to total harvest quantities, to seasonal or daily collection times and areas, and to individuals involved in primary collection. In Limbi village in southern Gujarat, households maintain historical 'first' rights to individual *mahua* trees (*Bassia latifolia*) and their flowers. If a family fails to claim its rights by burning the grass in the flower-fall area at the base of the tree, other families can legitimately claim rights to the flowers. In some cases, communities give individual households the collection rights to specified areas of the forest, a practice that may prove inequitable as the composition of the community changes. The community management group may also decide to give each household seasonal rights to harvest grasses. The arrangement may allow one member of each family to visit the forest daily during the harvest season, or the management group may limit quantities, such as one cartload of grass per family, a specified number of fuelwood headloads, or a certain number of poles per household.

Where autonomous community groups have developed their own rules, they are primarily governed by principles of need-based legitimacy, equity, and sustainability. In some places, if availability of timber trees is limited, cutting is permitted only after the management committee has first verified the need. The CI of Chargi in Bihar has decided that even the *Mukhiya*, or management committee, should obtain permission from the panchayat headman (*Sarpanch*) before cutting any trees.[26] In typical situations where availability of forest produce is limited, extraction is often regulated by designating days or months when one or more individual per family is permitted to harvest. In other cases, user-fees and/or permits serve to regulate extraction and distribution. Fees may vary with the type of produce, its availability and with quantities harvested. Many CIs have devised rules that require extraction permits from the leadership. Each member of the HRMS of Lohgarh in Haryana must obtain a permit from the society's president to collect one free bhabbar grass headload entitled to each household for the season. For fodder grasses, a seasonal harvesting fee (*dati*) per collector is charged. The dati rate is often lowered or waived, through consensual agreement by HRMS, for the poorest villagers or those likely to harvest less frequently. Where user-fees are charged, non-members may be permitted to harvest or collect non-timber forest products (NTFPs) by paying a higher rate. As another type of regulatory mechanism, some self-initiated groups in Orissa permit only women household members to collect NTFPs from the protected forest. In contrast to permits for individual members to collect products when desired, some CIs opt to have specific seasonal NTFPs harvested in one discrete period for subsequent distribution and/or sale to members. Funds raised through user-fees or sale are used to cover the CI's administrative expenses or applied to other community activities.

Under any one of a range of benefit-sharing systems, group members must feel that the division and distribution mechanism is equitable according to relevant, measurable criteria. This is the case whether the system is based on greatest need, prior rights, or voluntary labour contribution to protection activities, or a combination of these. If the system is perceived to be equitable and operating smoothly, the management committee or leadership will gain legitimacy and the respect and support of its community members. Alternatively, if the system is seen as biased and unfair, or corrupt and inefficient, the organization may lose credibility and ultimately collapse for lack of support.

Conflict Resolution Mechanisms

Every CI is likely to face situations of periodic conflict. A certain amount of conflict is inevitable as the CI's persuade its members to foresake some individual benefits or freedom for a larger, common goal. In the case of CIs engaged in forest management, potential conflict can fall into four categories: within the CI's membership; with neighbouring nonmembers; with other external commercial or industrial interests; and with the state, primarily the forest department. For its own sustainability, it is crucial that each CI evolves effective conflict resolution mechanisms. Older, well-established CIs with a trusted leadership are often able to resolve local conflicts internally. For conflicts with external interests or agencies, mediation by a neutral third party may be required. In new forest department — promoted CIs, intervention by forest department staff or NGOs may be useful to facilitate resolution of either intra- or intervillage conflicts.

Intravillage conflicts are more common in heterogenous villages with prominent class and caste differences and wide variations in the nature and types of forest dependencies of different subgroups. Clash of interests between a grazier subgroup desiring forest access and others desiring closure to facilitate regeneration, is a classic example of the type of predictable conflict that is difficult to resolve. Perceived inequity in the distribution of costs and benefits of forest closure, doubts about fiscal integrity, obstinacy of some members in accepting common rules, and suspicions that the leadership is unduly favouring its own vested party can all lead to intravillage conflicts.

On the other hand, conflicts with neighbouring villages often arise over boundaries, denial of forest access, or the usurpation of the rights of a weaker community by a more powerful one. With the forest department, conflicts can emerge over direct clear-felling of forests on which local villagers are highly dependent, or over the forest department's sale of timber rights to commercial or industrial interests. The goal of many self-initiated protection groups in Bihar and Orissa is, ironically, to protect their forests from the forest department. Where JFM programmes have been initiated by forest departments, conflicts with CIs may be generated over the imposition of inappropriate rules or demands, the refusal of the forest department to honour its commitments, the uncoordinated functioning of different wings of the agency, or the forest department's failure to understand or acknowledge the community's problems, priorities, or knowledge.

The most effective mechanism for the resolution of intervillage conflicts in areas with strong traditions of CIs is frequent and regular general body, or *gram sabha*, meetings. Most autonomous forest protection groups in Bihar organize such meetings, binding all members to attend them. Attendance is taken seriously — the penalty for nonparticipation is the loss of membership. Representative leaders from each *tola*, or hamlet, also attend. If a conflict cannot be resolved through open discussion, the leadership is also expected to step in. If the collective leadership is unable to arrive at a satisfactory decision, the responsibility passes over to one or more respected individuals whose decision is binding.

Intervillage conflicts may also be resolved through discussion and negotiation among the leaders of the concerned villages, particularly where forest department or other government presence is weak. In other areas where villagers are participating in forest management with encouragement from forest departments, field staff are often called upon to play a neutral role as facilitator. The JFM support team in Haryana has frequently been called upon by HRMS to assist them in resolving both intra- and inter-HRMS conflicts. Disputes over and contradictions in many of the initial assumptions on which Haryana's JFM programme was originally based have similarly been resolved with the help of a continuous dialogue between HRMS and the JFM support team.

The Haryana experience shows that conflict arbitration within and between HRMS has generally been easier to achieve than to ensure any consistency in decisions taken by the Haryana Forest Department. Constant political and institutional flux, including repeated transfers of forest department personnel, has resulted in the placement of individuals with widely differing attitudes, understanding, and priorities concerning JFM. This has disrupted the process of dialogue and confidence-building necessary for HRMS. The problem currently appears inherent in the structure of the forest department bureaucracy and remains an unresolved issue in promoting JFM in different states.

Conclusions

A review of Indian experiences with grassroots forest management reveals that effective community groups usually possess some of the following charactersitics: high forest dependency, perception of resource scarcity or outside threats, geographical proximity to the forest, prior or current formal or informal rights, presence of indigenous resource management institutions, traditional socio-religious forest values, and strong local leadership. Democratically representative and participatory community institutions of forest-dependent, resident villagers are often the most effective organizations for ensuring sustainable management of forests, as opposed to groups established through government directives and projects. Forest departments can benefit enormously in their efforts at improving the current degraded state of India's forests and their long-term management by soliciting joint partnerships with these small, often informal community institutions.

In contrast, the gram panchayat — which has usually been viewed by the government as *the* local institutional vehicle for developmental projects and resource management — has proved to be disappointing for a variety of reasons. In the gram panchayat, most decision-making powers are vested in the elected representatives, while voters have neither the power to recall ineffective or corrupt leaders, nor the authority to responsibly assist their representatives to function more effectively. The panchayat representatives, on the other hand, complain of lack of co-operation by their voters.

As a viable alternative, the majority of the autonomously formed forest protection groups in India have clearly defined the roles and responsibilities of both the leadership and regular members. In tribal areas particularly, these are rooted in strong traditions of collective action. It would help the forest department and others in JFM support roles to further study these indigenous, grassroots institutions in order to develop more sophisticated guidelines for new CIs forming in other areas.

To effectively perform their functions in JFM, community

resource management institutions must be founded on generic institutional principles, including commitments to equity, autonomy, and participatory decision-making. Given the vast diversity in local resource management traditions and capabilities, expansion of JFM across the country will need to be carefully conceived and closely monitored. The forest department will need to help and support local communities in developing their own capabilities and solutions to suit the diversity of their situations and concerns.

No centralized decision-making system can unilaterally evolve appropriate interventions, given the complexity of existing human-ecological relationships in India. Thousands of self-initiated local organizations in Bihar, Orissa, Karnataka, West Bengal, and other states have already demonstrated their ability to practice sustainable forest management based on the principles of equity and grassroots democracy. Instead of attempting to make these CIs conform to standardized, institutional forms imposed from above, the challenge for forest departments lies in listening to and learning from their potential allies in protection and management — the communities themselves. The wealth of experience, local wisdom, and diversity that the nation's rural communities represent will inform the forest department and other facilitators how best to support and strengthen locally-inspired initiatives while also helping to promote the spread of forest community empowerment and re-generation of the forest ecosystem.

Notes

Acknowledgments: The author has relied heavily on her personal experiences during thirteen years of involvement with the Haryana Forest Department's JFM programme. She has benefited greatly from the sharing of insights with friends and colleagues, particularly with Mr J.R. Gupta, co-ordinator of the Haryana JFM support team, and with other professionals working with JFM in other states. Special thanks are due to Dr Mark Poffenberger, Ms Betsy McGean, and Mr Jeffrey Campbell for their valuable comments and suggestions. Gratitude is also due to several village women and men of Haryana's HRMS who continue to provide the most pertinent insights to JFM — provided one remembers to listen and learn.

1. See, Ramachandra Guha, *The Unquiet Woods: Ecological Change and Peasant Resistance in the Himalaya* (Delhi: Oxford University Press; Berkeley: University of California Press, 1989), and M. Gadgil and R. Guha, *This Fissured Land: An Ecological History of India* (Delhi: Oxford University Press, 1992).
2. Ibid.
3. Agarwal Anil and Sunita Narain, *Towards Green Villages* (New Delhi: Center for Science and Environment, 1989).
4. Neera M. Singh and Kundan K. Singh, 'Forest Protection by Communities in Orissa: A New Green Revolution', *Forests, Trees, and People*, Newsletter no. 19, 1993; N.C. Saxena, 'Joint Forest Management: A New Development Bandwagon in India?' ODI, Rural Development Forestry Network Paper 14d, 1992.
5. M. Chakravarty-Kaul, 'Durability in Diversity: Two Centuries of Co-management in the Punjab Forests, An Institutional Analysis', draft paper for publication, mimeo, 1993.
6. S. Kant, Neera M. Singh, and Kundan K. Singh, 'Community-Based Forest Management Systems: Case Studies from Orissa' (New Delhi: IIFM, SIDA, and ISO/Swedforest, 1991).
7. See M. Chakravarty-Kaul, 'Durability in Diversity'.
8. M. Poffenberger, B. McGean, and K.K. Singh, 'Orissa Trip Report', memo to NSG members, 29 August 1992.
9. Personal communication from S. Palit, CCF, Social Forestry, West Bengal Forest Department, 11 June 1992.
10. Mark Poffenberger, 'The Resurgence of Community Forest Management: Case Studies from Eastern India' (prepared for the Liz Claiborne and Art Ortenberg Foundation Community-Based Conservation Workshop, Arlie, Virginia, 18–22 October 1993).
11. See S. Mehrotra and C. Kishore, *A Study of Voluntary Forest Protection in Chotanagpur, Bihar* (Bhopal: IIFM, 1990); See also Ramachandra Guha, *The Unquiet Woods* and M. Gadgil and R. Guha, *This Fissured Land*.
12. Centre for Science and Environment, *The State of India's Environment, 1982, A Citizen's Report* (New Delhi: CSE, 1982), p. 118.
13. S. Pati, R. Panda, and A. Rai, 'Comparative Assessment of Forest Protection by Communities' (paper presented at JFM workshop, Bhubaneswar, 28–29 May 1993).
14. National Commission on Self-Employed Women, *Shramshakti* (New Delhi,.1987).
15. See S. Kant, N.M. Singh, and K.K. Singh, 'Community-Based Forest Management Systems: Case Studies from Orissa'; See also S. Mehrotra and C. Kishore, *A Study of Voluntary Forest Protection in Chotanagpur, Bihar*.
16. Neera M. Singh and Kundan K. Singh, 'Forest Protection by Communities: The Balangir Experience', mimeo, 1992.
17. See Mehrotra and Kishore.

18. S. Pathan, N.J. Arul, and M. Poffenberger, 'Forest Protection Committees in Gujarat: Joint Management Initiatives', Sustainable Forest Management Working Paper Series, no. 7 (New Delhi: Ford Foundation, 1990), pp. 19–21.
19. See Mehrotra and Kishore.
20. See N.M. Singh and K.K. Singh, 'Forest Protection by Communities: The Balangir Experience'.
21. See S. Pathan, N.J. Arul and M. Poffenberger, 'Forest Protection Committees in Gujarat: Joint Management Initiatives', pp. 19–21.
22. V.K. Bahuguna, *Collective Resource Management: An Experience in Harda Forest Division* (Bhopal: RCWD, IIFM, 1992), p. 10.
23. Personal communication from B. M. S. Rathore, DFO, Harda Forest Division, Madhya Pradesh Forest Department, 17 July 1993.
24. Ibid.
25. See S. Mehrotra and C. Kishore.
26. Ibid.

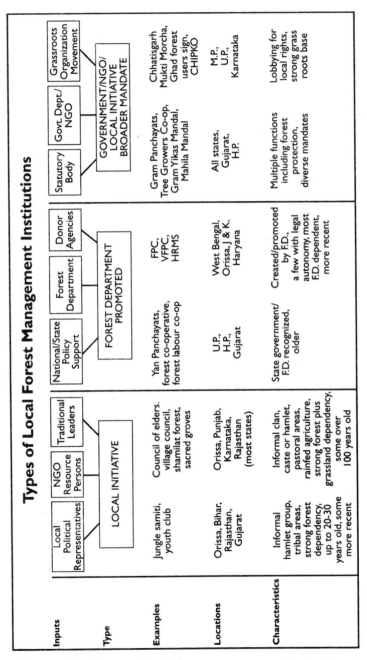

FIGURE 6.1 Types of local forest management institutions

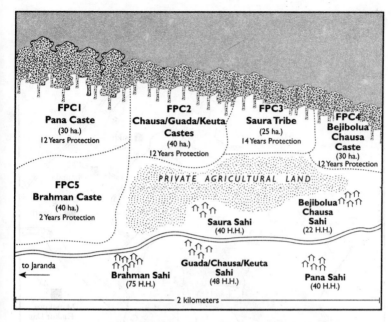

FIGURE 6.2 Social composition of forest protection groups
in Mahapada village, Orissa

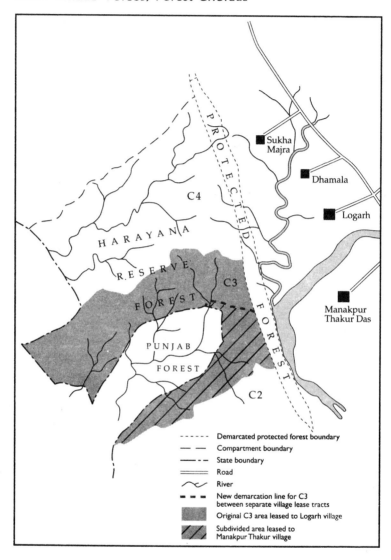

FIGURE 6.3 Forest areas protected by Mahapada hamlets

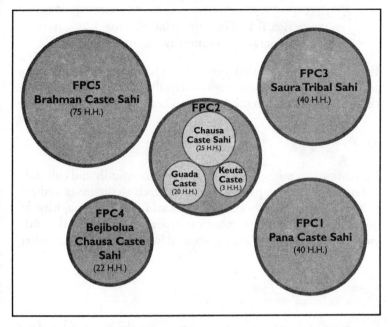

FIGURE 6.4 Dhamala forest beat:
Reallocation of forest compartment C3 to two villages in 1991

Box 6.1: Guiding Principles for
Democratic Community Institutions

Experience indicates that certain characteristics are closely associated with the effectiveness and stability of community institutions in managing forest resources. These include:

Viable social unit of organization

Institutions comprised of small, socio-economically and culturally homogenous groups tend better to reach a consensus and act cohesively as resource managers. In addition, the majority of members should have a common interest and similar dependencies on the resource, and be situated in proximity to each other and the forest.

Organizational norms and procedures

Community institutions must have the ability to formulate norms and procedures, whether formal or informal, to control resource use behavior. These must be based on the principle of equitable rights and responsibilities of all members, and be accepted by the majority. Desirable norms include keeping membership open for all resident adult women and men, assuring effective representation of all minority groups, and making a commitment to promote gender equity.

Accountability

The sustainability of community institutions will be jeopardized without effective mechanisms for ensuring accountability of individual members and the leadership. Mechanisms that help stabilize the organization include clear and accessible records of decisions and accounts, annual elections of formal representatives, participatory and open decision-making with full women's involvement, the general body's power to recall corrupt representatives, and penalties for violation of consensual rules.

Conflict resolution

Periodic conflicts between individual members or different interest groups are inevitable in any organization. Consequently, an effective community institution must maintain a variety of reliable conflict-resolution mechanisms. These could include arbitration through a strong and trusted leadership; access to impartial, respected outside individuals; or a monitoring agency such as the forest department, panchayat, council of elders, or the managing committee.

Autonomous status

For any community institution to function genuinely as the voice of its members, it must become an autonomous entity. For effective participation in a JFM partnership, its creation and dissolution should not be controlled by the forest department. Both parties must have an equal right to terminate the *agreement* between them, but not to disband the partner institution. Apart from entering into a formal JFM agreement with the forest department, the community institution should also enjoy an independent legal status.

Chapter 7

 Indian Forest Departments
in Transition

Subhabrata Palit

Indian forestry is at the crossroads today. Under mounting criticism for its inability to halt the loss of forest cover, a new generation of foresters is beginning to re-examine their role in managing these critical national resources. For nearly one hundred years, Indian forest management was shaped by colonial rulers whose goal was to strengthen government control over resources to ensure their continued commercial availability. Indian forest officers were trained in paramilitary traditions to implement colonial policies. This professional cadre was expected to rule a vast empire. As the state's authority extended, it undermined the traditional community rights and indigenous use-systems. Only recently have foresters and planners begun to realize that their role, of policing India's vast forest areas, was neither responding to the needs of nature nor to the nation's rural communities.

This chapter briefly examines Indian forestry traditions and suggests the types of management changes that may help to sustain India's forests as the nation moves into the twenty-first century. The current crisis in Indian forest management, as reflected in the rapid degradation of the nation's once-rich and diverse ecosystems, underscores the failure of past management and the need for a radical reversal in policy. Fortunately, rural people, tribal communities, non-governmental organizations (NGOs), and creative foresters are finding new ways to work together and reverse the deforestation process. This chapter explores changing attitudes, objectives, operating policies and procedures among forest agency staff, and the forging of new human relationships with rural communities that need to take place to move from conflict to collaboration.

Joint forest management provides a new basis for forming equitable alliances between agency staff and rural communities to achieve the common goal of preservation and continuing productivity of natural forests. Joint forest management lends forest departments the logistical support and motivation of millions of rural people whose survival depends on these resources. Forest departments can extend legitimacy to community management groups, supporting them politically, economically, and technically in their efforts to enhance productivity and improve their standard of living. Joint forest management is not only an attractive strategy for degraded forests but may also be the best option for protecting well-established productive forests and national parks. The transition from custodial to collaborative management has sweeping implications for policy, and for procedural, attitudinal, and technological reforms.

The Forester's Changing Perspective

During the mid-nineteenth century, the concept of 'scientific' forest management was promoted by the British colonial government. The commercialization of forestry required homogeneous, even aged crops that could be easily managed and harvested to optimize profits. Since emphasis was primarily placed on a narrow range of timber species, non-timber forest products were considered of secondary importance. Conventional forestry viewed the activities of resident communities as 'biotic interference', that must be minimized, if not altogether eliminated, to protect forest resources.

From an institutional perspective, the administrative infrastructure that was developed during the colonial rule proved grossly inadequate in the post-Independence era. This was especially true as the fast-changing political environment generated new leaders and policies, some of which began to seriously undermine bureaucratic authority. While newly independent India had adopted the entire forest legal codes of the colonial era, it was developing a new political system. In some cases, forest laws became unenforceable. People living on the forest fringes saw their resources exploited by politically supported business people. Even honest foresters had little authority to halt the abuse of forest resources. Ironically, both the state forest departments and forest-dependent communities suffered from the insecurity of forest resource tenure and access. Under such conditions, sustainable forest management became an impossible goal.

During the 1960s and 1970s, forest departments viewed rural communities as the primary impediment to successful protection. Communities, on the other hand, perceived forestry personnel as usurpers of their legitimate rights. As forests deteriorated and vital natural resources grew scarce, tension mounted on both sides. The more committed foresters envisioned their protection efforts as a temporary, generally ineffective attempt to halt a continuing and inevitable process of forest degradation. This frustrating situation alienated officers at all levels, leading to widespread corruption through illegal alliances between field staff, contractors, and local people at a lower level and among senior officers, politicians, and business people at higher levels.

As forest resources became increasingly depleted, some forest officers attempted to restrict logging. However, a growing number of rural people depended on illegally collected timber and fuelwood for the expanding energy requirements. A contractor–collector nexus emerged, draining forest resources and transferring them to urban centres. To break this nexus, market raids were conducted by many forest departments with the help of police who would seize the illicit forest products and store them at depots. Confrontation and conflicts were so intense that hundreds of Indian forest staff lost their lives in defending the forest from illegal extraction and overexploitation. Thousands of impoverished villagers, both men and women, were also killed, injured, or imprisoned. Throughout the period, foresters were often viewed by communities as officious and ruthless or as corrupt bullies, while many foresters viewed the villagers as devious, ignorant, and uncooperative. Throughout the 1960s and 1970s the gulf between foresters and rural people grew.

In the early 1980s, social forestry programmes were initiated as a positive step towards easing pressures on the declining state forests. The strategy was cautious, even conservative, in its effort to produce and distribute plantation resources to the people rather than involve them directly in the improved management of natural forests. The more successful social forestry plantations comprised eucalyptus raised as a cash crop, chiefly for the commercial pole market. Unfortunately, these plantations did little to solve local fodder and fuelwood scarcities and, consequently, did not alleviate pressure on the natural forests as intended.

Although attention and funds in the 1980s were channelled predominantly to social forestry programmes on private and designated community lands, millions of hectares of natural state forest land continued to be degraded. The financial resources of the administrative units managing natural forests, known as territorial circles, were often limited in contrast to the well-financed social forestry projects sponsored by foreign donor agencies. This imbalance created conflicts within some state forest departments and tended to make social forestry, with its large project budgets, more attractive to the staff.

Social forestry programmes, however, encouraged forest department staff to gain experience by working with village groups, initiating a process of staff reorientation. Yet, while foresters were beginning to work with village panchayats and community groups to establish woodlots, deforestation of India's natural forests continued at a rapid pace. Throughout the 1970s, commercial logging continued with leases frequently being provided to contractors at well below market prices. Poor villagers, seeing their neighbouring forests being commercially exploited, saw no reason why they should not gain at least a small share of the products. After the forest had been logged, poor rural families dependent on fuelwood headloading and livestock grazing entered the area and continued the disturbance, further suppressing regeneration. Consequently, disadvantaged communities living on the forest peripheries inadvertently became both agents and victims of the ongoing destructive pattern. This rapid devastation has finally led to a rethinking of community forestry strategies in national forest management policy.

While supportive state and national participatory management guidelines are being formulated, antagonistic traditions remain strong, and old attitudes are still firmly entrenched. Colonial objectives which aimed at maximizing government revenues or protecting forests by excluding local users no longer respond to national or local priorities. National policies that ensure ecosystems are preserved, watersheds are maintained, and subsistence requirements of hundreds of millions of rural Indians are met, need to be formulated. The need to define and implement policies that respond to national priorities while supporting India's tribal and forest people, is a challenge that Indian planners are faced with today.

Creating Supportive Staff Attitudes and Skills

Western systems of forest administration have been followed in most Indian states for more than a hundred years. The attitudes of forest officers have been shaped by training programmes and the supervision of their senior officers. The gradual shift toward joint forest management constitutes a major change in the minds of most foresters. Top-down channels of communication need to be replaced by an open dialogue between senior and junior staff, and community members. Hierarchically structured forest departments will need to build new professional and technical expertise, and skills in human interaction. More importantly, rural communities, once seen as encroachers and destroyers of forests, must now be viewed as colleagues and equal partners in management.

As management transformations are made, some forestry staff may experience a sense of insecurity, given the philosophical and operational reversals required. Support at the highest policy level is therefore necessary. Forest ministers, secretaries, and senior conservators will need to be seriously committed to the bureaucratic reorientation if it is to succeed. Principal chief conservators and their senior staff will require opportunities to explore and develop new management objectives and strategies to implement participatory programmes. They also require time to hold informal discussions with their mid-level and field staff to examine operational problems and opportunities that may arise during the transition to decentralized community control. Through this dialogue, senior officers can be reassured that organizational changes are within their control. They can examine the benefits of decentralization, supporting required actions rather than resisting them. If senior officers are supportive of management changes, junior staff will feel encouraged and reassured when implementing new policies and programmes.

Junior staff have different reasons to support joint forest

management (JFM). In the past, field staff have had to confront the inconsistencies of policies that mandated complete state control. In the field, a few guards were left to patrol vast tracts of forests used by thousands of villagers and illegal loggers. Field staff were often beaten and sometimes killed when they attempted to intervene. Formal empowerment of community groups as forest managers provides field staff with allies in management, making them collaborators in a common cause, rather than alienated agents of the state. Experience shows that with support from senior officers, some field staff can rapidly adapt themselves to co-operative management practices. While the attitude of many junior staff towards JFM programmes may be positive, information regarding new policies and procedures will need to be systematically transferred. Training and field monitoring, as well as assistance from NGOs, can help accelerate this process.

To implement JFM successfully, all staff, especially those in the field, will have to provide support to community groups, assisting them in formulating effective resource management systems. This will necessitate developing new skills in collaborative planning and integrated development activities. Foresters will require access to technical information in many unconventional fields not covered in previous training programmes. These new areas may include soil conservation, social forestry, non-timber forest products, horticulture, sericulture, and lac and mushroom cultivation. Senior staff will benefit from exposure to new planning processes, non-timber production systems, integrated forest ecosystem management, and new approaches to integrated watershed development. Field staff will require specific training in JFM policies and programmes. It is also necessary to transfer managerial and technical skills to communities to enable them to act more effectively as partners under joint management agreements. Training activities might include nursery care, planting of trees, cleaning and mulching, and coppicing, as well as methods to improve the profitability of processing and marketing non-timber forest products.

Forestry field staff can facilitate the transition to joint management by discussing management needs, community concerns, and priorities with various community groups. This will also help foresters restore their own credibility with rural communities and begin building co-operative working relationships. Foresters can also facilitate problem-solving by discussing forest management problems with NGOs, panchayats, local leaders, and villagers. Informal conversations with villagers will often encourage them to initiate protection activities. Ultimately, enduring collaborative management will be best ensured by open communication, trust, and honesty in interactions between foresters and communities.

Creating a Meritocracy

Joint forest management involves a working partnership between the forest department and participating communities. Forest departments will need to demonstrate a high degree of integrity to attract community groups to work collaboratively with them. Insincerity or dishonesty by departmental staff can jeopardize joint activities. In financial matters, foresters should disclose budgetary availability and expenditure information to community management groups. Forest departments will need to be accountable to their community partners, while village forest management groups will have to meet their responsibilities towards the forest department. Forest department staff, who acted unilaterally in the past, may find it difficult to accept this new relationship.

In West Bengal, foresters were made accountable to community groups as early as 1984, when the Panchayati Raj (local governance) system was adopted. New government guidelines required the forest department to keep panchayats informed of work programmes, financial outlays, and labour recruitment. Initially, there was considerable resentment among the staff against this new form of accountability. One senior officer who was visiting a forest district during the period recalls being questioned by a local forest ranger who was having problems working with the local panchayat *samiti* (council). He complained that the local panchayat officials were demanding to be informed of all forestry activities in his range and had asked to see all work estimates and vouchers. The ranger noted that while he was trained to carry out the commands of his superiors in the department, he was not accountable to outsiders. The senior officer tried to explain that under the new guidelines, the forest department was obligated to report to the local government. He explained to the ranger that the people's representatives had the right to ask for detailed expenditures incurred for public purposes. This case illustrates the types of conflicts that the forest department staff face as they are asked to become more accountable to the public, after more than a century of freedom to operate.

With the introduction of JFM and microplans, departmental accountability to local communities for public spending will grow. Local governments and community groups will want to scrutinize expenditures related to microplans, which will also help control corruption within forest departments. However, forestry staff will inevitably resist these new procedures. Orientation courses can help prepare agency personnel psychologically for changes in their authority and responsibilities. The existing parameters for assessing staff performance also need to be reviewed. In the past, dedicated officers who took initiatives to help communities resolve conflicts over forest resources and improve management systems usually worked in isolation. Sometimes these efforts were noted in the individual's service record, but often officers were criticized for taking risks, and for appearing to be overly ambitious. Forest department staff who attempted to respond to community problems could also come into conflict with the unscrupulous practices of local elites or dishonest colleagues, possibly causing the officer career problems. In most cases, no policy existed to support these rather informal efforts to build co-operative action between the department and forest villages. When the officer was transferred from the area after three to five years, the agreements were usually forgotten and the divide between the department and the communities would reappear. Often, in an unsupportive bureaucratic environment that provided little reward for dedication, honesty, and social service, forest department staff had to decide whether to submit to and participate in corrupt practices or pursue their work as honest professionals and face ostracism.

New policies for staff promotions, which provide professional incentives for meritorious service, need to be formulated and implemented. The transition to a bureaucracy where promotion is based on merit rather than seniority alone will encourage a new generation of foresters, allowing the most effective and dynamic officers to rise to positions of leadership. This would facilitate in bringing about a qualitative improvement in the functioning of forest bureaucracies and the management of forests. Staff policy changes would also demonstrate the seriousness government attaches to JFM.

Changing the Structure and Functions of Forest Departments

In India, there are currently more than 120,000 forest department staff involved in management activities. Many of these individuals are burdened by administrative duties and have little or no time for field work. Yet, they are solely responsible for safeguarding nearly 23 per cent of the nation's land area and arbitrating access for the millions of rural people involved in using forest lands for collection and grazing purposes. Senior planners realize that staff levels are grossly inadequate to deal with the mounting pressures on India's natural forests. The adoption of JFM provides an opportunity to substantially reduce policing duties by the staff by transferring this role to community forest protection groups. However, while taking some custodial pressure off field staff, foresters will need to allocate increasing amounts of time to help organize and provide technical support to community management groups.

With community groups emerging as local forest managers, new administrative issues arise regarding ways for forest agencies to interact effectively with thousands of village management units. In West Bengal, a field officer may need to interact with thirty or more forest protection committees (FPCs) operating in his area (beat). Experience indicates that in order to maintain good communications during the first few years after group formation, the officer may need to contact each FPC at least once a week. Given the limitations on his mobility, the field officer may have difficulty in effectively handling more than six groups. In addition to routine communication, field staff may need to assist communities in developing microplans for managing and developing their forest areas, which often vary considerably according to the local site conditions and community priorities.

In many parts of India, where communities are forming forest management groups, these organizations often consist of twenty to hundred or more households. In south-west Bengal alone, more than 2,200 FPCs exist. Close co-operation between FPCs and forestry field staff requires frequent interactions. Six to ten FPCs per beat officer may be the optimal workload. To achieve such intensive coverage, however, beats, ranges and territorial divisions may have to be reduced to smaller units. Current experience indicates that community management groups are beginning to cluster around small but contiguous patches of forest. Although it may make sense to place a field officer in charge of each significant patch or cluster of FPCs, this would immediately require creation of many new posts. Since the government is not in a position to hire new staff, it may be better to adjust the workload either through a structural reorganization of forest departments or by obtaining NGO help to cover areas under joint management.

Frequently, territorial staff are overextended under the combined burden of patrolling large areas and handling routine administrative duties. Technical divisions within the forest department, like soil conservation or social forestry, are well staffed, but have no direct responsibility for territorial management. Staff from these technical units could be made directly responsible for supporting community resource management groups in specified territorial areas, in addition to their specialized work. This will remove the existing overlap between territorial and technical divisions, generating additional personnel who can be redeployed to reduce beats and ranges to a more manageable size. Purulia district in West Bengal has been successful with staff redeployment. The original territorial division forest officer's (DFO) area was divided into three sections, with management responsibilities shared with the soil conservation division chief and the social forestry DFO who were already assigned to the area. Some state forest departments may prefer eventually to reintegrate some technical circles, like social forestry, into the territorial circles.

These organizational changes require a redefinition of job assignments, an equitable distribution of workload, and a structural reorganization of forest departments. Since the reorganization affects not only the DFOs but all their subordinate staff, it must also provide opportunities for guards, beat officers, and rangers to participate in planning the adjustments in assignments. This would also allow for establishing new systems of bottom-up and lateral communication. Divisional field staff who have already adjusted to the new operating procedures could share their knowledge with colleagues and senior officers in other units, thereby resulting in more effective forest department operations.

Local governance institutions and planning processes are being decentralized under the Panchayati Raj system being introduced in many states of India. In some cases, it would be strategic to make the revised beat and range boundaries cojurisdictional with the Anchal Panchayats, or block-level government territories. This will facilitate the integration of forestry sector activities with the larger developmental planning and monitoring process. At the same time, the updating of the Indian Government's national map survey during the 1990s may provide an opportunity to formally demarcate joint forest management areas.

NGOs can also respond to the deficiencies in government infrastructure, particularly in training, participatory research, localized planning, and programme implementation. NGOs that can assist with these activities need to be identified both at the district and state levels. State-level NGOs can organize training programmes for the senior-level staff and carry out supportive diagnostic studies. At the district level they can help in organizing orientation and training programmes, in holding microplanning exercises, in disseminating market information, and in resolving disputes.

Demarcation of FPC Lands and Registration

A fundamental step in empowering communities as local managers of public forest lands requires neighbouring villages and the forest department to agree on the territory for which responsibilities will be delegated. Communities that protect natural forests usually assume control of areas based on historical use-patterns and their informal usufruct rights which are often already recognized by neighbouring villages. New FPCs that are in the process of forming sometimes consult neighbouring groups to seek their approval in an attempt to avoid future conflicts. Despite these efforts, conflicts over rights do occur and usufruct areas sometimes overlap, resulting in disputes and mismanagement.

Forest departments need to ensure that all affected communities agree before formally demarcating tracts under community management. This will help clarify protection and use rights and responsibilities for the territory under JFM. Forest department staff can facilitate meetings among community groups to assist them in reaching a consensus regarding the division of forest management responsibilities. Discussions regarding demarcation should be held as soon as a committee is formed. In areas where informal groups are already protecting forests, the forest department may only need to ensure that neighbouring villages agree with the current territorial division of management responsibilities.

Since external boundaries of the larger forest tracts are already demarcated by the department, the primary objective is the demarcation of internal boundaries. In reserve forests, the boundaries for a block or a compartment could form the management territory for an FPC; however, sometimes new subdivision lines will need to be drawn to better conform to local use patterns and traditional usufruct rights. After an FPC has been provisionally formed and the land area has been demarcated, forestry field staff must register the new committee to ensure their rights to forest products and to give them legal legitimacy. Registration should clarify the membership of the group and provide details of the land area managed, including a map, to minimize future litigations.

Forest Management Planning

Working plans have traditionally been the primary guidelines for state-managed forests in India. The divisional working plans are normally drawn up for a twenty-year period and legally dictate what development and management activities are approved in the area. Under JFM, however, collaborative management agreements between forest protection committees and the forest department are based on microplans. While working plans tend to emphasize timber planting and felling cycles, microplans usually stress a more diverse range of non-timber forest products (NTFP) enhancement programmes, soil and water conservation works, and other development programmes.

Microplans also provide formal documentation identifying the community management groups and the forest territory for which they are responsible. Microplan development strategies should reflect the available resources in the areas, investment resources, people's needs and aspirations, and an agreed formula for benefit sharing. They are formulated in consultation with community members who determine development priorities within the available options. Microplans also serve as a memorandum of understanding between the forest department and the community and ensure the continuity of a programme. Currently, all activities included in the microplan need to conform to the working plan prescriptions, and to the Indian Forest and Conservation Acts. In eastern Indian states, where thousands of FPCs operate, equal numbers of microplans need to be formulated and monitored. To facilitate supervision and planning, it may be necessary to computerize microplans. Formats for this purpose have already been developed in a few states.

In many cases, outdated working plans need to be revised to reflect changing objectives, placing greater emphasis on integrated forest ecosystem management, environmental protection, and community needs and controls, while de-emphasizing timber production. As working plans are revised, microplans can be used as building blocks to reformulate beat, range, division, and ultimately state forest management strategies. Microplans should be prepared for a five-year period to synchronize with the state development programmes.

In areas where JFM systems are widely adopted, working plans will become an aggregate of microplans. In such cases, they will have little relationship to earlier working plans with their general prescriptions for large areas. Rather, a new analytic and administrative framework for supporting decentralized microplans will need to be created, detailing regeneration methods, development strategies, sharing agreements, and silvicultural operations. Microplans will highlight specific local needs and community development priorities within a broader legal and strategic forest management programme, for a region based on community priorities and on changing ecological and economic conditions.

The success of JFM will depend on a long-term commitment to improving forest resources through effective forest department-community collaboration. Five-year microplans can provide a basis for this type of sustained co-operative action. Adoption of a microplan would also involve a commitment by the government. Government failure to respond to the agreement would be viewed by the communities as a breach of trust and may lead to a breakdown of the system and consequent destruction of forests. Therefore, availability of a dependable budget is essential for JFM support actions. Long-term financial commitments to the programme should also allow for flexibility and timely responsiveness to meet the diverse needs of forest communities.

Experience indicates that fiscal flexibility is more important in effectively supporting community activities than the volume of funding. Although JFM and natural regeneration often require only modest financial support, the forest department's ability to respond quickly to community needs is a necessity. While state forest departments can gain access to funds from bilateral and multilateral donor agencies, JFM projects can be initiated with modest amounts of existing funds. In West Bengal, project support from the International Development Association was only acquired after several thousand villages began protecting forests throughout the south-western part of the state, demonstrating the effectiveness of participatory management in regenerating degraded sal forests. During the first ten years of programme expansion, very modest budgetary support was used. It may be that the absence of excessive capital support allowed the programme to evolve on its own, without target-driven project pressures, ensuring that FPCs were responding to local interests, rather than donor or agency project priorities.

In-service Training and Formal Forestry Education

There is a growing recognition that forestry course curricula urgently need to be revised in many training institutions throughout the country. Most forestry training programmes continue to focus on traditional methods of timber stand management and park protection; little or no attention is given to the social and economic contexts in which forest management must take place. Graduating foresters are unprepared to deal with forest-dependent communities, non-timber forest production systems, or with the challenges of integrated forest ecosystem management.

Reforms in training curricula require both central and state government support. New training programmes for trainers and incentives for professional forestry educators must be developed to stress the importance of these changes. Systematic attempts to restructure forestry education must be carried out simultaneously from the prestigious Indira Gandhi National Forest Academy, through the state forest service colleges, rangers colleges, and forest guard's schools nationwide.

Syllabus revision committees should be carefully constituted to bring in experienced individuals from within and outside forest departments. Opportunities should be created for state-level committees to revise training programmes and to exchange ideas on content and teaching strategies. New curricula should provide greater opportunities for students to discuss new policies and programmes, including participatory management initiatives. Greater emphasis should also be placed on field exercises and discussions with community management groups.

Monitoring and Research Support

Joint forest management represents a major departure from the custodial management systems of the past. This transformation of management systems could lead to the rapid regeneration of large areas of degraded forests; however, failure could further disturb or complicate existing management problems. To ensure that early experiences from collaborative management are used to improve programme implementation and to identify problems before they become too disruptive, it is important to establish an independent monitoring and evaluation unit.

Apart from ensuring a steady flow of information from field experiences, it is critical that this information is used to inform decision-making. The establishment of a programme working group — comprised of senior officers, field staff, and NGO groups — can provide a forum to review field experiences. Working groups will need to meet regularly to review and follow-up on programme strategies and problems. Working group agenda, minutes from meetings, and follow-up reports can provide a data base to document the evolution of the programme. In this way, information generated through the Management Information System (MIS) can actually be used for management purposes.

A transition to collaborative forest management has broad implications for research priorities, methods, and analytical requirements. Traditional forestry research has focused primarily on silvicultural aspects of important timber trees, including their growth and the quality of wood they produce. In the future, research priorities need to shift to studying the larger structure and function of forest ecosystems. NTFPs, which were termed as minor forest products in the past and had a limited role in research, have gained much greater importance. Foresters, villagers, and others will need additional information regarding ways to enhance productivity and income through culturing a wide range of herb, shrub, climber, and tree species, as well as strategies to improve processing and marketing. Research will also be needed to monitor the effects of policies and programmes on community institutions. Currently, the economically more valuable NTFPs are being over-exploited, with some species facing severe depletion or even extinction. New culturing and enrichment planting strategies need to be designed to increase the productivity of desirable species present in natural forest ecosystems.

Forestry research institutions need to reassess their priorities to respond to the emerging needs of communities managing natural forests for a range of products. NTFPs and ecological studies of forest disturbance and natural regeneration patterns need to be given high priority. Long-term ecological monitoring studies could be initiated through forest management and research agencies, while short-term diagnostic studies of forest product processing and marketing systems could be conducted by NGOs.

Managing Production Forests Through JFM

Deforestation and the failure of forests to regenerate after harvesting as prescribed in the working plans, have led to decreasing logging projections. In addition, the Forest Conservation Act of 1980 and its subsequent amendment in 1988 have virtually banned clear-felling of natural forests. These actions have substantially decreased logging activities and diminished associated employment opportunities. While demands for timber for industrial needs will continue to exist, it is likely that timber production will be increasingly limited to plantation areas outside forests. A shift away from timber production in natural forests, however, provides excellent opportunities to explore how non-timber forest products might be harvested on a sustained basis.

Conventional timber production, because of its longer rotation periods, may not be the best management system in areas where maximizing employment and income-generating opportunities are priorities. Collection, processing, and marketing of NTFPs can, in some cases, provide substantial income to the community, although the volume availability of different products will vary. A recent study of a regenerating sal forest in south-west Bengal indicated that income from NTFPs was three times higher than that from timber during a ten-year rotation. Natural forest production systems, due to their great diversity of species, provide a steady and more immediate flow of benefits to forest communities, unlike single-species plantations. Additional income and employment can be created through implementing support activities such as sericulture, mushroom cultivation, lac cultivation, orchid propagation, basket-making, weaving, pisciculture, and bee-keeping. Other government departments and agencies — agriculture, minor irrigation, scheduled tribes, horticulture, sericulture, and panchayats — may also be able to assist community management groups. FPCs provide the government with opportunities to assist highly motivated rural organizations to better use existing resources.

Joint forest management has wide applicability in India's production forests. Most government resolutions indicate that this approach is primarily required for degraded forests. In absolute terms, most of India's forests suffer from some form of degradation. As a consequence, community management should not be limited to badly degraded ecosystems. Good standing forests would also benefit by community-imposed access controls, which reduce biotic interference, slowing or halting the degradation process, while allowing regeneration to occur. Waiting for forests to become extremely degraded before allowing communities to protect them makes little management sense.

Community groups managing natural forests often need to allocate considerable labour to patrolling their area and protecting it against grazing and cutting. Community members may also be forced to forgo opportunities to cut fuelwood, obtain fodder for their livestock, and gather other forest products, especially during the early years of forest regeneration. For the lowest income groups in the village, some of whom may depend heavily on forest products as a source of subsistence and cash, ceasing exploitation can have serious economic implications. The forest department should provide them with the means to compensate for the opportunity costs and for their subsistence.

Community forest management organizations also provide villagers improved bargaining power in dealing with NTFP processing and marketing agencies. Strategies to assist NTFP collectors to aggregate their products and eliminate market-chain intermediaries need to be developed. Forest departments need to explore better ways to ensure that good prices are available for NTFP collectors. Attempts to establish marketing structures have often reinforced lower prices, rather than improving them. Diagnostic studies are required to assess the effectiveness of current government marketing support agencies to see if they actually bring better returns to collectors. Where they fail to do so, alternative marketing support strategies need to be developed.

Since forest departments require time to develop new capacities to support village management groups, it is important that these agencies introduce new JFM programmes gradually, as their ability and effectiveness grows. Targets for forming forest management groups may put undue pressures on field staff and communities. If the staff has not been properly trained and new procedures instituted, initiating complex, well-funded JFM activities may be counterproductive. At the same time, forest departments should encourage communities and field staff to co-operate wherever possible.

Managing Protected Areas through JFM

Experiences with JFM systems in protected areas of India and elsewhere in Asia indicate that participatory management may be effective in helping to conserve biodiversity. Protected areas represent some of India's best forest land and deserve the most effective management systems available. India's protected area consists of 70 national parks and 411 wildlife sanctuaries covering 20 per cent of India's forest area and almost 4 per cent of the total geographic area. Most of these protected areas are surrounded by numerous communities, which live in villages within the park and sanctuary boundaries. Protected area and wildlife policies have generally required closing forests to local communities, placing broad restrictions on the collection of fuelwood and fodder and on grazing.

Communities living on the periphery of sanctuaries and national parks are exposed to hazardous destruction of life and property by elephants, rhinos, and bison. Attempts to resettle people outside the boundaries of protected areas have generally disrupted community organizations and their economic activities.

The alienation and dislocation of forest communities from protected areas have frequently resulted in conflicts. A 1989 report[1] on national parks and sanctuaries reveals confrontations between the forest department and villagers over access to natural resources. From 1979 to 1984, fifty-one violent clashes occurred in national parks and sixty-six in sanctuaries. Unless people living on the peripheries of the protected areas are involved in their management, neither its wildlife nor habitat can be preserved. The Wildlife Protection Act will need to be amended to provide opportunities for communities to play a role in protecting the nation's biological resources. At the same time, usufruct benefits can be extended to community management groups, especially in buffer areas where degraded forests may be regenerated.

Conclusion

It is increasingly evident that without the commitment and co-operation of forest-dependent communities, the forests of India will remain in jeopardy. A concerted effort must now be made to regenerate India's forest ecosystems. Forest departments, communities, NGOs, research and academic institutions, and other government agencies must join forces and mobilize resources and co-operation from all possible sources.

Through collaborative efforts, forests can be protected, regulated, and ultimately conserved. Field evidence increasingly shows that participatory or joint forest management is the only practical method that can counter deforestation and ensure sustainable forestry in overpopulated, resource-scarce countries. Joint forest management need not be restricted only to degraded forest lands. The strategy also has the potential to respond to forest management needs of protected areas such as sanctuaries, national parks, and biosphere reserves where traditional communities have historically relied on forests for survival.

To stabilize India's forest ecosystems, foresters will need to make a tremendous commitment to reforming their institutions and management systems. These changes need to be informed by strategic, well-funded research and training programmes. To make these changes, foresters will require strong political and administrative support from the government. New forest management strategies will be needed to restore the health and productivity of degraded ecosystems. Joint forest management provides a unique opportunity and challenge to professional foresters to bring a renewed effectiveness and commitment to social service to their institutions.

In several states in India, a promising beginning has been made. In recent years, Indian foresters have become more aware of the need to transform their institutions, informed by bitter but revealing experiences. For the new century, our target must be a forest that is environmentally sound, has a sustainable production base, responds to the needs of local people, and is managed by them as a stable and healthy ecosystem. The challenge to the forester of the twenty-first century is to learn to manage these values efficiently.

Notes

1. Ashish Kothari, Pratibha Pande, Shekhar Singh, and Dilnavaj Variara, *Management of National Parks and Sanctuaries in India: A Status Report* (New Delhi: Indian Institute of Public Administration, 1989).

Chapter 8

 Learning to Learn: Training
and Gender Sensitization in
Indian Forest Departments

Betsy McGean ■ S.B. Roy ■ Mitali Chatterjee

Joint forest management provides an opportunity for Indian forest departments and rural communities to move from a history of opposition and conflict towards co-operative action in management. This transition assumes the development of new communication channels, attitudes, and relationships that can help overcome barriers of hierarchy, caste, class, ethnicity, and gender. Methods and strategies to facilitate this reformation of forest management systems are currently being used in India. This chapter reviews emerging strategies to facilitate learning, including the use of working groups and participatory appraisal methods, interactive training programmes, and gender-sensitization activities.

A growing number of state forest departments have formed working groups, comprised of senior officers, NGO representatives, and researchers, to discuss forest management problems and solutions at the policy and field level. New operational methods such as micro-planning are being used to involve villagers in identifying priorities for forest protection and development. These decentralized planning techniques rely on participatory diagnostic tools that enable villagers to discuss forest use practices and changes in the resource over time and location. Participatory rural appraisal (PRA) methodologies for forest planning have been developed by teams working in West Bengal, Orissa, Haryana, Gujarat, and Karnataka and have proved to be effective in providing other foresters, NGOs, and researchers with a clear picture of community–forest relationships, problems, and opportunities.[1] At the same time, data collection and analysis provide rural people with an opportunity to review their own resource use practices, identifying areas where changes in local management could be beneficial.

Dozens of training programmes for both senior foresters and field staff have also been conducted by NGOs in several states to explore ways to work more effectively with local forest management groups. This chapter highlights some of the most important experiential lessons in reorienting attitudes toward communities and women, and modifying field duties and practices in the nation's forest departments. Before examining these experiences, however, it is useful to examine negative attitudes held by foresters, villagers, women, and NGOs that often constrain communication and the establishment of trust and co-operative working relationships.

Creating a Learning Environment: The Role of the Working Group

New strategies for forest restoration and management are being evolved by some of the nation's poorest tribals and lower-level forestry field staff. While a select group of senior officers, researchers, and NGO leaders are increasingly providing support to community initiatives, solutions are being found from within the forests and the villages. It is essential that urban-based elites who control funding, policies, and planning, listen to and learn from the experiences of the rural people.

Reorientation programmes, which help inform forest department staff, planners, NGOs, and other involved groups regarding the goals and resource management strategies being adopted by local communities, are an essential component in facilitating the transition to collaborative management. Yet, conventional training will need to be complemented with other learning mechanisms to accelerate organizational change and facilitate the upward flow of information from the field. These would include regular working groups, informal meetings and briefings, workshops, field partner identifications, PRAs, focused diagnostic systems research, and monitoring.

Working groups are increasingly being used by Asian forest departments that are involved in innovative participatory management programmes. Their function is to act as a communication forum where parties meet regularly to discuss progress in management programmes. Ideally, a working group should allow a select group of senior foresters, field staff, researchers, and NGO representatives to freely discuss emerging operational problems and opportunities related to new management systems. To be effective, working groups need to meet regularly and be supplied with a steady flow of field information, which highlights difficulties encountered by the staff and village management groups, and the emerging strategies to respond to these constraints, allowing the working group to ·develop supportive policies and programmes. The regular meetings, usually held every four to six weeks, provide opportunities for field staff to inform their seniors, and for NGOs to provide new information about which the agency may be unaware, allowing them to modify and adjust project components (see Figure 8.1).

During working group meetings, monthly activities are outlined, and responsibilities assigned. These assignments help keep staff accountable and responsible to the group as a whole. Experience from West Bengal, Haryana, and Gujarat, indicates that a working group may benefit from an outside facilitator, especially in its early years of inception. The facilitator can help convene meetings, set agendas, and follow-up on implementation activities.

Another strategy, which has proved effective in West Bengal, involves less structured, informal meetings between the staff of participating NGOs and senior officers. In this case, a facilitating NGO meets with the chief conservator of forests. These meetings are not intended as a forum for advice or consultation but to exchange ideas, field practises, and learning. Such informal discussions across the table often prove to be a highly effective mode of information-sharing, and ultimately, of reorientation and training. This is particularly important for those senior officers who may be reluctant to participate in formal training activities.

Accelerating Learning through Workshops and Seminars

As state programmes unfold, workshops for foresters, NGOs, and researchers from different parts of the country can be held to exchange information regarding experiences with joint forest management (JFM). These workshops may include a discussion of state strategies, case studies, production and processing of non-timber forest products (NTFPs), gender-related technical and production issues, forest protection regulations, micro-management planning, and extension and training needs. The workshop as a forum for information-sharing provides an excellent educational foundation for officers struggling with similar problems. Over the past five years, two national workshops have been convened by the central Ministry of Forests for state ministers, senior forestry officials, NGOs, researchers, and donors. These workshops have helped assess the national status of the JFM programme and underscored some of the more pragmatic operational strategies for improving the sustainability of forest-management systems under JFM. In 1990, the first of these national workshops resulted in the creation of a national research network committed to pursuing some of the major ecological, economic, and institutional issues confronting JFM.

At the community level, some NGOs and forest departments are exploring new ways to identify key village partners in an effort to transform local forest management practices. This has involved identifying a strategic village leader, a women's organization with particular strength and interest in forest protection, or another village-level institution, which could play a valuable role in comprising or advising a forest protection committee. In Orissa, local foresters near the Simlipal National Park have allied themselves with a network of approximately eighty villages that co-ordinate forest protection in the region. While these villages have concentrated their management activities in the reserved and protected forests near their communities over the past decade, the Orissa Forest Department is working with village leaders to explore how the communities might collaborate with them in extending protection to the park area. In Karnataka, scientists and NGOs are encouraging the forest department to work with local schools and colleges, as well as with community-managed sacred groves, to document forest biodiversity and ensure its sustainable management.

In south-eastern Gujarat, the forest department staff organized a ten-day march (*pada yatra*) through more than twenty tribal villages to publicise the need for greater forest protection and discuss the new JFM initiative with community members. Each night a fair (*mela*) was held in the village which also provided an

opportunity for discussion. The march proved to be very successful in opening new communication channels, reducing conflicts,
and providing avenues to establish new joint management agreements with communities in the area. Three years after the march,
several hundred villages in the region had begun working with the
forest department to protect degraded forests.[2]

Many strategies and mechanisms are needed to open communications between India's vast rural community and state forest
agencies. Working groups provide opportunities for learning and
upward-flow of information to senior officers, while bringing them
into sustained interaction with NGOs and researchers who share
their concerns for better forest management. Informed by diagnostic case studies, monitoring reports, and discussions with field staff
and village leaders, working groups can accelerate learning and generate new policies and programmes based on the latest knowledge.

Developing Effective Training Programmes

Over the past five to ten years, a number of states in India have
embarked on experiments with joint forest management, with
differing degrees of success. They have generated important information and lessons in such areas as village resource management
institutions, participatory micro-planning, and agency reorientation and training. Despite the progress in participatory management in such well-documented villages as Arabari in West Bengal
and Sukomajri in Haryana, interviews with field staff continue to
underscore the need for reforms within the forest department. The
attitudes and concerns of beat officers in Haryana illustrate some
of the most urgent issues facing forest departments across India.[3]
These include the need for reorientation and training, including
social skill-building, attitude adjustment, incentives, recognition,
and new modes of communication within and outside the institution. The value of forestry field officers as the operational informants and key counterparts to the community to guide the emerging
JFM programme is highlighted below.

Haryana Field Staff Perspectives on Training Needs

Throughout the 1970s and 1980s, while social forestry field

programmes were generously funded, little attention was given to modernising forestry training institutions or to updating their curricula, to reflect changing management priorities. Without guidance through formal and in-service training, forestry staff were unable to respond to the changing needs of the forest departments or to the needs of participating forest communities. In Haryana, for example, despite the state forest department's achievements in facilitating the creation of several dozen registered Hill Resource Management Societies (HRMS), field staff have received little formal training to build their capacity to work effectively with villagers. Interviews with field staff revealed that one beat officer, who had been intimately involved with the programme for more than a decade, had yet to participate in any structured HRMS orientation activity. His colleague did attend a six-month course, but the contents of that course directly contradicted the objectives and operational strategies of JFM. The curriculum reportedly encouraged field staff to adopt an intimidating, authoritarian posture when dealing with villagers and prepared them only for traditional custodial work in the forest.

The beat officers emphasize the importance of new reorientation and training programmes in helping them to understand the conceptual underpinnings, and learn new skills to carry out JFM. They recommend that curriculum be carefully designed to respond to emerging management systems and that it be mandatory for both formal induction training and in-service programmes for existing staff.

Significantly, the suggestions offered by the beat officers on the most important components of a new training curriculum focus primarily on interpersonal skill-building and much less on technical or procedural competence. These features include sensitive, non-threatening methods of approaching villagers, building rapport, and communicating respect to elders; organizational strategies to facilitate the development of community management institutions and protection systems; effective methods of conflict-mediation in dealing with village factionalism and violations; and process documentation research techniques. Suggestions for building new technical expertise involved participatory sketch-mapping, village-level data collection, production monitoring, and soil/water conservation methods. The beat officers recommended that a one-week course be introduced, combining classroom work with frequent field visits and exercises so that trainees can absorb new information, apply their new skills in field situations, and analyse immediate feedback in a controlled environment. One officer said, 'I have learned the most by sitting with the District Forest Officer (DFO) and the NGO facilitator, while they were organizing JFM meetings and talking to the villagers. I am observing their new attitudes and behaviour towards the villagers, including recognizing their ideas and needs, and building respect and rapport with them.'

Committed, but struggling over the past four to five years to implement participatory management against significant institutional barriers, the officers interviewed feel that the JFM programme offers major improvements over the more traditional system of custodial state control. These field advisers have formulated creative proposals and strategies to address a full range of issues — facilitating the creation of Hill Resource Management Societies (HRMS) in areas where villagers are hesitant and poorly informed; setting limits of ten per cent annual increases for bhabbar (fibre grass) leases; creating a federation of HRMS to better negotiate leases and oversee agreements and tax exemptions. Yet a fundamental problem is that, as currently structured, the forest department provides no institutionalized forum for bottom-up communication for the field staff to contribute their experience to the senior management, thus obstructing the systematic channelling of grassroots field-learning — a critical foundation for an evolving programme. Individuals best suited to provide a continuous feedback on the day-to-day constraints and opportunities for successful field implementation are marginalized from decision-making. More often than not, field staff who have tried to communicate their problems and needs have been ignored and made to feel inconsequential by senior officers. However, the beat officers are perhaps more strategically positioned and practically informed than any other agency individuals involved in JFM. If encouraged, they can serve as the eyes and ears of the forest department's programme-monitoring, midcourse adjustments, and ultimate success in supporting community management.

The beat officers of Haryana have contributed many ideas regarding ways to reform the current forest department procedures to remove institutional barriers and facilitate the integration of joint forest management (JFM). They recognize that their roles

1. Men in uniforms with a paramilitary orientation continue to create social distance between foresters and villagers (M. Poffenberger).

2. In Bhainsara hamlet (*phalan*), 20 kilometres north of Udaipur, in Rajasthan, the Bh
hamlet chief says 'we have protected all the forest land on the far side of the ridge f
ten years and all the trees and grasses are growing well. Two neighbouring villages a
ready to join us. Sickle fees are charged for each family that cuts grass in the forest a
we use the money to hire watchers from our village who guard the forest from outside
who may try to fell our trees' (M. Poffenberger).

3. Using sketch maps Sikh, Jat and Banjara villagers work with J.R. Gupta of the Haryana Forest Department to clarfiy management responsibilities in Shivalik hill forests using sketch maps. (M.Poffenberger).

4. A Vasava tribal woman in southern Gujarat shows foresters and village leaders, on a ground sketch map made of local materials, where she travels in the forest to collect mahua, gum, fodder and fuelwood (M. Poffenberger).

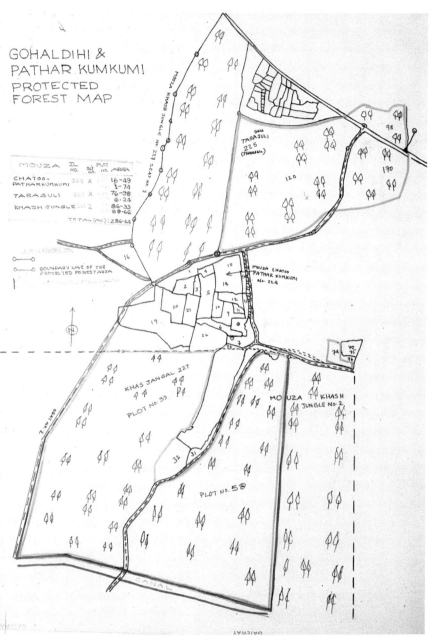

This map was commissioned by a forest protection committee in West Bengal. The villagers hired the local tax collector to draw the map to better clarify the boundaries of their protected forest (M. Poffenberger).

6. The removal of root stock is a final blow in the impoverishment of the for
ecosystem. Without these roots, topsoil erosion will accelerate and rapid regenerat
possibilities are lost (M. Poffenberger).

7. Bhil women from Bhainsara break for lunch after planting trees on degraded forest lands. Tens of millions of Indian women are primary forest users. New means must be found to bring them into management decision making (M. Poffenberger).

8. In arid Western Rajasthan, the indigenous Bishnoi people have survived for centuries, tapping an extensive knowledge of the desert scrub and forest environment. Their continued presence in the desert is ensured only by their sustainable management

will change substantially under the new JFM programme; however, they feel that this will be balanced by the assistance offered by the community in protection and management activities. In some cases, HRMS members have informed the forest department of illegal forest activities and assisted the beat officers in apprehending offenders. In addition, the number of police cases related to forest offences has declined, significantly reducing the time the field staff would have to spend in court. This observation has been confirmed by forestry field staff in other Indian states, as well as in Indonesia, Thailand, and the Philippines where JFM is being practiced. Most important, beat officers report that their primary incentive to encourage the establishment of JFM systems is the vastly improved relations that they can cultivate with the community. Villagers frequently grow to treat the field officers as trusted friends, inviting them to social functions and consulting them as advisers on other civic affairs. This historic transformation of a relationship — from conflict to co-operation between the community and the forest agency — offers a new promise for securing the requisite institutional foundation necessary to ensure that forest resources are managed in a sustainable manner.

The beat officers, however, also point out the realities of their opportunity costs in pursuing JFM. Informal payments, which are forgone in the JFM programme, still exist with fellow officers who decide to pursue more traditional but illegal practices outside the accountability of the community. For this reason, additional incentives are needed to further motivate those already practising JFM, while simultaneously convincing others that apart from money, professional gains may be more rewarding in the long run.

Another disincentive to field personnel is the practice of frequent transfers and new postings without any warning or consultation. This issue is often raised by beat officers as being counterproductive to their efforts at nurturing a relationship of continuity, trust, and self-confidence with community groups. One staff member reported that he was transferred four times during an eighteen-month period. Frequent transfers are typically based on political interference rather than programmatic need, and can undo months or years of progress in building supportive relationships between the forest department and villagers. The beat officers believe that posting durations for JFM should be no less than three years and that assignments should be made based on merit, motivation, performance, and programme requirements.

A major professional incentive, as cited by young officers, is promotion based on merit and performance, as opposed to seniority, politics, or pay-offs — a much more common pattern. A solid meritocracy can be further fortified by special financial increments based on achievements in working with communities. A reinvigorated incentive package might include awards, certificates, training and travel opportunities, special recognition at ceremonies and in the media, and compensatory housing and educational assistance for field officers assigned to remote areas.

Beat officers believe that one of the most essential but often missing aspect of their redefined roles in JFM is recognition from their superiors. Some field staff fear that senior officers are more likely to treat critically a lower-level officer who pursues a new, alternative form of participatory forest management. Within the forest department, the creation of new working relationships between senior and junior officers is as essential as it is between foresters and villagers. New communication channels can be forged only if senior officers take the lead and insist that field staff are given ample and regular opportunities to express freely their concerns and ideas for improvement. Senior officers will need to recognize the ultimate futility of protecting India's vast forests through policing alone and to communicate unequivocally the necessity of involving communities as equal partners in management.

New Training Approaches for Forest Department Staff

Numerous NGOs have been experimenting with training and reorientation methodologies in the JFM programme over the past five years. Through research and training network exchanges in India, a general consensus regarding some of the most promising training strategies in the classroom and field is emerging. A sequential training programme for foresters, which creates both a conducive learning environment and introduces new concepts and materials gradually, allows participants to open themselves to new information and experiences. While all levels of forestry officers in the institutional hierarchy need opportunities to discuss issues and implications of transition to JFM, this is especially important for lower-level field officers who play a pivotal community-liaison role in the programme. Yet, some NGOs specializing in training emphasize the need to hold meetings initially with senior officers. They contend that only when the agency elite both understand and accept the concepts and strategies involved in JFM, will field officers have the political and operational support they need to modify their behaviour and implement the new programme.

Some senior officers in the Orissa Forest Department contend that it is important that the staff not only understand and support what communities are attempting to do in their local management initiatives, but also have a thorough knowledge of the techniques for supporting village efforts. They note that until significant progress is made not only in shifting staff attitudes, but in building joint management capacities within the department, it is better for the department not to interfere or confuse village-based forest management activities.[4] Consequently, it is important for forest departments to develop and implement a thorough staff training and reorientation programme prior to launching strategies to influence spontaneous community efforts toward forest protection.

For a successful training session it is important to create an informal, non-threatening, learning environment. This is accomplished by establishing a classroom in which all participants, regardless of hierarchy, sit in a circle. The Indian Institute for Biosocial Research and Development, an NGO that has pioneered training programmes for forest departments, has found that for most participating forest department staff, it is often the first time that they have been together where no distinction between junior and senior status is acknowledged. Everyone in the room is a student, and learning is naturally interactive, adaptive, and synergistic. The next series of interactions involves the 'ice-breaking' stage of the training, whereby participants become more open, communicative, and attentive. Posturing and gesturing is analysed to allow senior officers to learn how to be more relaxed and accessible when dealing with junior staff. Trainers lead the participants through participatory agenda-setting processes, which can be used for both staff and community meetings.

As the group begins to communicate informally, the participants examine their values and attitudes towards each other and the forest communities. Instructors present practical case studies that explore issues such as gender in forestry, indigenous community knowledge, social justice, civic responsibility, and public service. These discussions help the forest department staff to become more aware of their past attitudes towards each other and rural communities. Skill-building exercises are held to help the staff with techniques for process documentation, interpersonal communication, participatory problem-solving, transactional analysis, conflict resolution, and consensual decision-making. The skill-building exercises involve role-playing and case study analysis. This is followed by field-work in which participatory rural appraisal (PRA) methods are used to facilitate communication and learning between foresters and villagers.

In developing the forest department's capacity to work with villagers to formulate locally relevant micro-level management plans, it is important to build PRA skills. PRAs involve the rapid generation of specific, local-resource information through direct interactions with, and thorough involvement of, numerous key informants and village residents. The validity of collected data is then cross-checked through other sources. The PRA process helps to establish greater rapport between the forest department and the community, and to generate community-derived information that can help inform management planning and decision-making.

In initial encounters between forest department staff and villagers, responses of communities may not necessarily be positive. Given the history of conflict, rural people may continue to distrust the overtures of government officials. Mandatory formalities and introductions during community gatherings limit the time and atmosphere needed to elicit good, unbiased information. The ideas put across by the forestry officers about JFM are often unfamiliar. Villagers may be wary, based on their experiences with poorly planned projects, where agency officials came to provide employment. Several meetings may be required to convince the villagers that the forest department is sincere when it proposes a partnership based on people's participation and shared rights and responsibilities to forest resources. To diminish hierarchical divisions, forest department staff are encouraged to shed their official status by sharing a *charpoi*, or sitting on the floor, with the villagers. Perseverance, patience, and continuous follow-up by forestry staff are essential to cultivating communication channels and cementing a new relationship with the community.

A training exercise with the forest officers in Sarugarh, north Bengal, illuminates key issues including common misunderstandings, importance of relevant information, and possible processes for supporting community-centred JFM. In this case, the NGO facilitator (Institute for Biosocial Research and Development [IBRAD]) began by attempting to illuminate local issues, including specific management problems and conflicts. Earlier interviews with field officers in Sarugarh revealed that forest department staff perceived the villagers to be unsupportive of the JFM programme. The foresters assumed the community was upset by the lack of tangible programme incentives in the form of fuel, fodder, and employment. To better assess the validity of these assumptions, a three-day diagnostic PRA was held involving community members and forestry staff. The PRA revealed that local residents mistrusted the forest department, believing that the agency had not delivered on earlier commitments to provide employment for the community, to issue free identity cards to collect forest products, and generally to treat the villagers with more respect. The community also felt that JFM would further impede its ability to collect forest products. To resolve this dispute, IBRAD offered to assist in organizing a meeting to exchange perspectives.

A group exercise was held to sensitize forest officers and expose them to the villagers' views. During the initial role-playing exercise, some officers acted the part of the villagers, allowing others to practice listening to complaints without interrupting them, and developing ways to express appreciation and sympathy in response. Rather than rejecting the communities' more critical comments, the foresters learned how to respond with humility, accepting the villagers' criticisms rather than defending the forest department's position.

Soon after the role-playing exercise, a follow-up meeting was held in the village allowing officers the opportunity to experiment with their new listening-skills. Sitting on the ground under a tree, at the same physical level with the villagers, the training participants introduced themselves and expressed their interest in learning about community perspectives of local forest resource management. In an effort to cultivate a sense of trust, the officers admitted to the villagers that they had failed to protect the forests and that their best chance to sustain forests would be through full community participation, protection, and moral support.

This ice-breaking approach by the forest department trainees in Sarugarh proved effective in encouraging community openness and laid the groundwork for free discussion. During the meeting, villagers were relaxed and voiced their own perspectives, criticisms, and frustrations with the forest department. The forest officers listened intently, allowing the villagers to speak freely. Citing their own constraints and frustrations in attempting to reverse forest degradation, the officers suggested the idea of working together on a joint mapping exercise to delineate village forest tracts and access routes. Despite initial reluctance, the villagers began contributing to mapping the forest area. For the first time, community members began working co-operatively as equals with the forest department.

Feedback from several groups of officers, involved in participatory mapping, indicated its effectiveness in enlisting participation. Walking around the village as participant observers, the foresters developed a better understanding of the community and the villagers' lifestyle. Although the initial attitude of the villagers was predictably guarded, these barriers broke down when they became seriously engrossed in the interactions of PRA mapping. When the community's forest protection committee was marked on the map by a forestry officer, it surprised and pleased the villagers that their group was recognized. By illustrating the broad range of natural resources of the village on the map, the people became more aware of their collective resources, power to delineate boundaries, and take responsibility for forest protection and regeneration. Because of their daily interactions with and knowledge about forest and water resources, as well as their ability to identify the locations of detailed physical landmarks, both women and children were natural participants in the mapping exercise as well. This set an important precedent for their continued involvement. Once completed, the map served as a tool to stimulate ongoing discussions about resource management issues, eventually leading to an exercise involving ranking problems and priorities by the community. As an outcome of the interactive process of map-making, villagers gained confidence in their ability to work with the foresters and to commit themselves to future activities.

Over time, the feelings of mutual distrust between the community and foresters in Sarugarh were transformed as a new understanding between user groups emerged. The forest department was surprised to learn that their earlier preconceptions about the community's demands had been inaccurate. In fact, the villagers had not expected gifts of materials, money, or even immediate fuel and fodder handouts in exchange for their involvement. Instead they were seeking a relationship based on co-operation, openness, and trust with the forest department before making a commitment to participate in JFM. Similarly, the villagers learned that forestry officers were more understanding and less bureaucratic than they had earlier perceived. They found the foresters sincere in their request for the villagers' co-operation as partners in managing the forests and in their commitment to share forest products with them.

In short, training programmes for junior and senior foresters, as well as meetings that encourage exchanges between foresters and villagers, have proven to be effective in breaking down negative attitudes, reducing conflicts, opening communication channels, and providing a basis for a dialogue leading to better management agreements. The process through which these changes occur is complex. Over the past five years, the training and reorientation programmes conducted by IBRAD indicate that learning to be a good 'facilitator' requires time. Initially, foresters doubted the capacity of the NGO staff to provide training. IBRAD staff had limited knowledge of forest management issues and conflicts when the programme began. As the trainers learned more from the foresters and communities regarding management problems, they too became more effective in facilitating discussions and in helping to bring greater focus to critical issues. IBRAD has accelerated its own learning process through written and video documentation, analysing training and discussion sessions. IBRAD is acquiring practical knowledge and skills, as well as monitoring feedback concerning its reorientation, gender sensitization, and training work. Each new training programme incorporates previous experiences to improve its effectiveness.

IBRAD has demonstrated that if NGOs operate in a sensitive, diplomatic, and skilled manner, they can play a pivotal role in assisting with the reorientation and training of the forest department. A growing number of state forest departments across India are requesting training assistance from NGOs. Facilitators may also be based within forest departments, universities, or other government agencies. Ultimately, the job of the facilitator is not to 'train' people in the conventional sense but to create opportunities for the diverse groups involved in forest-use and management to overcome their conflicts, communicate effectively with one another, and forge viable management agreements.

Women in the Forest Department: Gender Sensitization Processes

Historically, Indian women have been the primary users of forest resources. Assuming major responsibility for the family's health and nutrition and serving as the primary collectors, producers, and processors of food for the household, women have long played an important role in the use and management of forest lands. With government land classification and eroding communal land systems, privatization, women's voice in management has diminished. As tribal forest cultures were influenced by Hindu groups, norms and values, which also undermined the role of women in communal resource decision-making, were gradually adopted. Patrilineal patterns of ownership and inheritance have not allowed women to own land and have undermined their authority over its care and management. Legislation has failed to establish full gender equality in law, much less in practice. Further, women have not been fully aware of this steady disempowerment process and the loss of their customary resource rights.

In countries throughout the world, India included, women as a subgroup comprise a disproportionate percentage of the poor. Employment opportunities available to women are often among the lowest in terms of wages generated on a per hour basis. In India, because such forest-based activities as the collection and processing of fodder, fuelwood, fibre grasses, bamboo, and reeds usually generate low hourly-income, they tend to be the domain of women. With the increase in forest degradation and fall in the productivity level of these disturbed ecosystems, the incomes of poor rural women are disproportionately affected. Due to their high subsistence dependence on the standing forest for such household requirements as fuelwood, fodder, medicines, fruits, and other foods, the quality of life for women and their families is also negatively affected. Most resource schemes, including social forestry programmes over the past two decades, have failed to recognize the diversity and significance of women's forest-resource needs; instead, they have emphasized cash flows with men in mind, assuming that benefits would trickle down from the male household-head.

Attempts to introduce gender awareness and equity into JFM programmes are hindered by broader patterns of discrimination found in society at large. Since women's participation in many sectors of the society is constrained by historical, cultural, and socio-political factors, questions arise regarding the effectiveness and speed with which their active involvement can be achieved in the forest management sector. Resistance to women's involvement in forest management can be strong among male foresters and village men, limiting opportunities for women. The involvement of women in forest management is not simply a matter of greater social justice and gender equity. In India, women are often the primary forest-user group, and without their formal involvement in management, agreements over sustainable use, access controls, and strategic development will be incomplete and, probably, not optimally effective. Given the complexity and challenge of achieving gender equality and justice in any sector of society, there is an urgent need to experiment boldly with different strategies across a diverse human-ecological context in order to understand what actually works in motivating, supporting, and sustaining women's involvement in forest management programmes. The economic, class, and caste variations in the position of Indian women as a subgroup need better documentation and analysis as do women's daily needs and roles vis-à-vis forest use.

Short-term JFM project objectives may need to target the poorest forest-dependent women with technical packages; however, ultimately programmes should treat women not only as beneficiaries but as primary actors, allies, and influential decision-makers in forest protection and management. Although the theory underlying gender issues in forestry has been widely elaborated, there are few well-documented case studies of field strategies that illuminate the opportunities and constraints involving women. More field investigations and process documentation are essential before women's full participation can be appropriately valued and institutionalized at the village level and within the forest department bureaucracy.

Increasing women's meaningful participation will require an approach that ultimately aims to elevate their status, allowing their entrance into the male domain of resource planning and management. Some state forest departments have attempted to bridge the gender gap and expand village women's roles in forest planning and implementation by recruiting more women at the field-level of the service. Given India's conservative rural traditions, male forest department staff have faced severe problems interacting with rural women. Gender segregation (*purdah*) and women's shyness have so restricted access to women that even the most committed male forest department officer has difficulty establishing effective communication, let alone more complex working relationships needed to implement JFM activities. As a consequence, a growing number of foresters and community development specialists agree that women forestry staff will be far more effective than their male counterparts in working with village women. Women in Orissa and other states have also expressed a strong desire for female extension workers to assist them with forestry-related projects. The following case explores the Orissa Forest Department's attempts to integrate women foresters within the agency.

Women and the Orissa Forest Department

Beginning in 1983, the Orissa Social Forestry Programme, supported bilaterally by the Swedish International Development Authority, made a concerted effort to hire recruit women as 'Lady Social Forestry Supervisors' (LSFSs) and 'Lady Village Forestry Workers (LVFWs)'. The primary objective of appointing women field staff has been to improve the department's ability to reach out to rural women and their children, and to mobilize their participation. The Haryana Forest Department has made similar efforts to recruit women village forestry workers into the agency, and more specifically into its JFM programme. A review of lessons learned from the Orissa and Haryana cases underscores some of the operational issues facing forest departments.

Women's Recruitment

Interviews with male and female field staff in the Orissa and

Haryana forest departments provide useful insights into gender problems confronting the agency professionals working with community groups. Most families supported their wives and daughters who sought jobs with the forest department, though in a few cases they feared for their security and reputation. Most of the applicants knew little regarding the nature of forestry jobs, or the roles they were expected to play, at the time of their interviews and acceptance. The women recruits tended to come from lower middle-class backgrounds and had compelling economic needs at home. At the same time, most of the female staff also said that they were interested in village service and community interaction, work satisfaction, improved status, and gaining greater economic independence through a job. By the third year of the project, the Orissa Forest Department had a core of forty young female village forestry workers, mostly high school graduates, who, at least in theory, could concentrate on gender issues related to social forestry programmes.

By 1990, the Orissa project was able to recruit only fifty-three female social forestry supervisors — representing 20 per cent of its target — and thirteen female village forestry workers, or 25 per cent of its goal. While noble in principle, hiring-targets may be impractical, and often counterproductive, if the forest department is not psychologically and operationally reoriented and ready to absorb women. In Orissa, the recruitment criteria and process were deemed inappropriate, as they were based on traditional territorial forestry positions although the nature of the field jobs for female social forestry supervisors and village forestry workers is vastly different. Instead of specifically tailoring the positions for female staff, job descriptions were identical for both male and female village forestry recruits. Furthermore, vague terms of references for the positions, rendered rank, authority level, and avenues for promotion ambiguous and elusive.

Orientation and Training

Although female staff in the Orissa social forestry programme are considered key elements in establishing communications with villagers, this function was not reflected in their training or orientation. The most important aspects of the job, namely extension education strategies and communication skills, were neglected in the curriculum. The entire training programme was limited to a thirty-day classroom course. During their orientation, the female village-forestry workers were neither adequately briefed on their specialized roles, nor were they given any particular sense of importance concerning their positions as women in relation to the longer-term gender-specific objectives of the project. As a result, the major reason for recruiting women was lost. Although some female field workers were placed in office jobs, others were made supervisors of plantation-forestry activities. This involved meeting physical targets for nursery and village woodlot establishments under unrealistic time·pressures. As a result very little attention could be paid to communication, motivation, and extension activities for village women. The one exception was in Sambalpur range, where local women were temporarily employed under an alternative arrangement as community motivators.

Emulating their more experienced male counterparts, female village forestry workers began to approach the more easily accessible male populations of the community, neglecting the village women. In the Haryana programme, new women recruits were often given assignments up to three years without any initial training. The system was so structured that only those exposed to training courses had any chance for future promotions. Hence, a high rate of female attrition was automatically introduced into the process.

It was also reported that during their training in Orissa, female social forestry supervisors felt physically and intellectually isolated from their male counterparts and the outside world. As one female trainee described, 'We were just like inmates of a jail', and another recruit explained, 'Our instructors wanted to play it safe — minimum interaction with the male trainees, restrict our going to the men's hostel — even our meals are sent here.' More importantly, although the social forestry supervisors were supposed to adopt a supervisory, managerial role and take charge of the range, they reported that during training any self-initiative or creativity was discouraged; instead, passivity and docility were rewarded.

Women's Experiences on the Job

Experiences in employing female field staff indicate that the absence of a supportive work culture and living environment has limited women's ability to perform their field jobs successfully. These problems were reflected in interviews with female forestry staff in both Orissa and Haryana.[5] Perhaps most crucial, a major 'barrier of acceptance' seemed to exist in the attitudes of many male superiors or managerial staff towards female field-recruits. This attitude towards women staff within the agency was detrimental to their performance both within and outside the department. Male project personnel did not typically perceive equivalent-level female and male staff as equals. For this reason, women were often discriminated against in promotions and as-signments. Some senior officers did not agree with the donor agency's policy to encourage the recruitment of female field staff, nor did they think it important to develop new department capacities to work directly with village women in the project. Instead, male personnel resented what they perceived to be an unfair recruiting process that allowed all women interviewees in Orissa to be hired, and that arbitrary project quotas for women had to be filled as a prior condition to receive foreign funds. Some male personnel also felt that women were incompetent, or capable of good work but unreliable. Others felt that women were willing to work but faced too many domestic and/or social obstacles: 'The ladies can work in cities and towns; in Delhi a lot of women are working. But in rural areas it is very difficult for them. The [village forestry worker] job is not meant for them.' In Haryana, one senior forester expressed a common belief that female staff 'always come to me crying with their babies' and that they expect special treatment.

As field extension workers trying to effect change, women staff in Orissa have faced serious institutional constraints. Since the budget was centrally controlled and provided no separate line of authority, there was no flexibility for independence or creativity in terms of programme development. For example, all publicity materials have been designed and produced through a central distribution centre, preventing the development of location-specific programmes or products. With no budget for extension activities at the range or beat level, there has been little incentive or opportunity to plan or implement new activities. In villages, with low literacy rates, visually-oriented extension materials are essential to communicate the goals and terms of new JFM programmes; yet very few educational-items have been available, and those too only on a random dispersal basis. As one female social forestry supervisor-trainee lamented, 'What is the use of studying about extension methods? Neither can we help design them, nor do we have enough when needed.' Not only are materials inadequate, but the extension programme has also suffered from a lack of clearly defined strategies to achieve project objectives.

The most widely voiced concern among the female village forestry workers has been the security problem of travelling alone after dark, which inhibits their ability to conduct evening meetings. As a consequence, most night meetings are conducted by male foresters, limiting women staff's input and influence in terms of community decision-making, sustained participation, and personal commitment to the project. Simultaneously, it has handicapped women's capacity to develop an independent professional identity and to interact informally in the community by building rapport and trust with both village women and men.

Another problem voiced by women was the inadequate travel allowance, and lack of vehicles. Committed staff report that they have to use their own funds to meet the costs incurred during field visits, thus limiting mobility. In addition, living conditions for female village forestry workers and social forestry supervisors were found to be poor. Housing was both inadequate and often located in places too secluded to meet security, medical, or material needs. Requirements of painstaking paperwork have been further disincentives.

Although designed as a field position, a sizable amount of paperwork was also required of the Orissa social forestry project supervisor; yet no clerical nor adequate office support or infrastructure was available. Hence, long field tours by female village forestry workers would often lead to severe administrative backlog. As a consequence, the social forestry supervisor would need to enlist a female village forestry worker to assist her with record-keeping, taking away five to seven days per month from the worker's own fieldwork.

Frequent transfers undermined the crucial cultivation of the community–forest department relationship. After months of building rapport and developing collaborative management strategies with village women and the larger community, a transfer could demoralize staff, and the momentum towards a more enduring JFM agreement could be lost. 'Every time you are transferred, you're back to square one, starting all over again. Establishing an environment of mutual trust is very difficult and time-consuming,' explained a female social forestry supervisor. Most of Haryana's efforts focused on recruiting women to play extension roles. However, with few exceptions, due to the prevalent stereotype among the senior officers that women were unsuited for 'tough' fieldwork and desired office jobs, they were transferred, without consultation, to new office assignments. This occurred even where women were demonstrating effective skills in field-level community organizing. Such new postings were often located far away from the resident village, requiring long and difficult commuting. Transfers often occurred with no explanation for termination, or description of the new assignment, further demoralizing the affected personnel.

While endeavours to enlist women field staff into the agency have encountered many problems, certain positive results have emerged. Acceptance by male superiors in the department may still be the largest obstacle for lower-level women field staff; yet the wider social acceptability of female village forestry workers has been consistently high in both Orissa and Haryana. In the village, women field workers gain prestige and respect for being employed in important community work. Within the family, employment brings improved status for the female village forestry workers and social forestry supervisors, often giving them increased decision-making influence. 'Who used to know who I am? Now they [the community] call me "Didi" and ask for my advice on various issues', proudly explains a female village forestry worker. 'I am contributing to my family income, so it has become natural on my father's part to ask for my decision', reported another worker.

Despite the limited institutional support available to female staff, women who conducted extension work in both Orissa and Haryana consistently proved their dedication, capability, and competence. The female village forestry workers in Orissa were considered by villagers as more consistent and 'softer' in their approach, and more efficient and effective in conveying the message of social forestry. According to one male superior, 'The LVFWs [lady village forestry workers] are liked more and accepted more easily by the villagers than the male VFWs [village forestry workers],' while others commented that communities perceived women as more sincere and honest than their male counterparts. Women field staff tended to command substantial respect within the village and in most interpersonal interactions. Women field staff in Orissa did report more difficulties in dealing with revenue authorities than their male counterparts. Women staff were particularly effective as motivators for co-operative protection (e.g. social fencing) of village woodlots, resulting in excellent plantations — some evaluated as the best. in the state. Case studies suggest that female field officers were particularly effective in facilitating community management activities when their presence was sustained in a village over an extended period. Research also indicates that village women may be most responsive and their participation the strongest when female field officers are able to meet them regularly.

Both Orissa's social forestry project and Haryana's experiences in employing women field staff shed light on many generic issues facing forest departments in their efforts to recruit women. Despite numerous problems, women forestry extension workers are demonstrating their effectiveness in the field. Given certain changes and improved opportunities, they could potentially have a much more powerful and positive impact on gender motivation and on the participatory aspects of the programme. Female field staff should be encouraged to target disadvantaged women, not only as an entry point to the village community, but as a high priority group.

In order to accommodate the unique and specialized roles for women field workers, a substantial restructuring of job descriptions, recruitment process, orientation and training, field priorities, infrastructure, and incentive system will be necessary. The major emphasis for female personnel must shift away from office and plantation work, and focus primarily on extension towards village-women as originally envisioned. Most crucial to the process of change, however, will be nurturing new, supportive attitudes among male counterparts and superiors towards women staff. Because it involves the empowerment of women, behavioural reorientation will require sensitive trainers, support from the top, time, and consistent follow-up programmes.

Summary

As India moves into the twenty-first century, new forest management strategies are evolving in response to changing demographic, environmental, and ecological conditions and needs. These new systems will need to heal old conflicts between the state and rural people, allowing co-operative action to emerge. They must confront gender biases that are deeply rooted in Indian society, to encourage women to play a formal role in forest management. Finally, the new management systems will need to break down status barriers among foresters, allowing field staff to contribute to the development of a more honest and effective working environment.

Over the next decade, well-conceived and executed in-service and formal training programmes will need to be implemented for thousands of foresters, community leaders, and NGO staff to develop greater capacity to establish joint forest management systems. Greater emphasis will need to be placed on the training of trainers and assisting the nation's many ranger colleges, NGO training centres, universities, and management institutes to establish relevant programmes with appropriate teaching materials. Staff recruitment, training, and field support can no longer be relegated to a position of marginal importance. A serious political commitment must be made by policy-makers, donors, and senior forest administrators to give priority to developing a new generation of foresters committed to the regeneration of India's degraded forests.

Thousands of women will need to be inducted into the Indian Forest Service and the state cadres. This phase presents an immense challenge for recruiting and training. Furthermore, the organizational environment of forest agencies should be reoriented to allow women to participate equally with their male counterparts. Working groups, diagnostic studies, new monitoring systems, and feedback loops that enable emerging experiences to be channelled into policy-making will transform these institutions, making them accountable to their staff and the public that they serve.

Notes

1. A two-volume set of PRA field manuals entitled *Diagnostic Tools for Supporting Joint Forest Management Systems* was published by the Society for the Promotion of Wastelands Development and is available from them at Shriram Bharatiya Kala Kendra Building, 1 Copernicus Marg, New Delhi, 110001.

2. R.S. Pathan, N.J. Arul, and Mark Poffenberger, 'Forest Protection Committees in Gujarat: Joint Management Initiatives', Sustainable Forest Management Working Paper series, no. 7 (New Delhi: Ford Foundation, 1990).

3. S.B. Roy and M. Poffenberger, 3 September 1991 memo on 'Discussions with Field Staff RE: Attitudes Toward HRMS Programme and Suggestions for Improvements'; Madhu Sarin, 1 May 1990 memo on 'Workshop with ROs and BOs of Naringarh, Panchkula, Raipur Rani, and Morni Ranges of Pinjore-Morni Division.'

4. Personal communication from Orissa Forest Department staff, September 1992.

5. See SIDA/IIFO/Swedforest report, 'Orissa's Female Villager Forestry Workers and Social Forestry Supervisors' (New Delhi, 1989). Also of interest — B. McGean and Rita Ray, 'Women's Involvement in Social Forestry: Mid-Term Evaluation' (Orissa: Swedforest/SIDA, 1991); Desmond Chaffey et al., 'An Evaluation of the SIDA-Supported Social Forestry Projects in Tamil Nadu and Orissa, India' (unpublished report, 1992); B. McGean, 26 October 1993 memo, 'Operational Strategies for Maximizing Women's Involvement in JFM: Inducting Women into the Forest Department.'

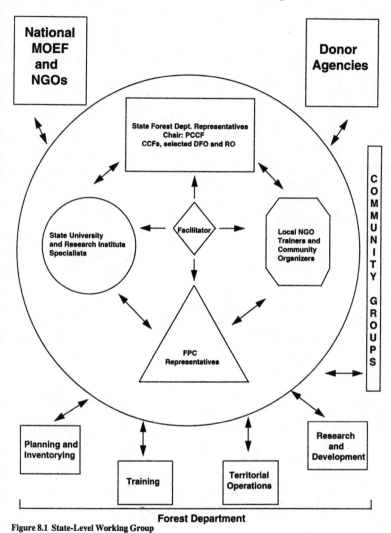

Figure 8.1 State-Level Working Group

FIGURE 8.1   State-level working group

Part III

Approaches to Sustainable Forest Management

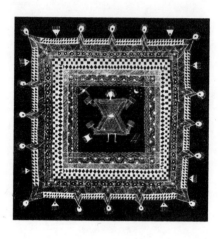

Chapter 9

Valuing the Forests

Mark Poffenberger

Ascribing values to forest resources is largely a subjective exercise. 'Worth' primarily reflects what is important in the eyes of the assessor. Market prices of many natural resources are non-existent or irregular. Where prices do exist, they may not reflect current or future scarcities, nor indicate whether exploitation rates reflect sustainability levels. Forest functions, even more than products, are extremely difficult, if not impossible, to value quantitatively.

Groups or individuals with similar socio-economic backgrounds and value systems tend to share a common perception of the value of particular products. For example, Santhal tribal women in south-west Bengal may rank forest foods and sal leaves for cottage industry plate-making as the most valuable products from the living forest. In another village, women may feel that the greatest benefit of the regenerating, secondary forest is the decrease in the seasonal migration of their husbands as a result of new, steady flows of subsistence forest products. Santhal children may rank high-nutrient forest fruit 'snacks' such as *ber* (*Zizyphus jujuba*) and *bel* (*Aegle marmelos*) as the most important forest products. In contrast, a business entrepreneur would more likely perceive the native sal (*Shorea robusta*) poles as the most highly-valued forest commodity. A conventional economist might agree with the businessman and place a market value on the poles rather than on the benefits resulting from reduced male out-migration or the nutritional value of forest-fruits. Local farmers may judge the forest's role in supporting long-term agricultural productivity as its most valuable function in the larger agro-ecosystem. Natural resource economists examining the upstream-downstream roles of watersheds may agree.

In the past, conventional economic analysis has underestimated the worth of the forest to the community. Economic anthropologists have started to calculate the value of forest products both in cash and in kind. In recent years, environmental functions of the forest ecosystems have also begun to be quantified in monetary terms by economists. With the growth of resource and environmental economics, more systematic methods have been developed to ascribe values to environmental externalities resulting from forest management policies. It is now possible, for example, to estimate the costs of logging an upper watershed in terms of the decreased life of a reservoir serving a hydroelectric project as a result of accelerated sedimentation. It is, however, still very difficult to make reasonably accurate calculations regarding the costs of flooding in Bangladesh on account of deforestation in Nepal.

Problems in valuation also arise from the paucity of information regarding non-timber forest products (NTFPs) and other benefits accruing to local communities over time. While exhaustive botanical studies have enumerated thousands of forest plant-species, these studies rarely consider use practices, processing, or marketing patterns. Further, ethnobotanical data are scattered. National macro-statistics are also often of questionable reliability. Since the establishment of a separate Ministry of Forestry in India in the late 1970s, most NTFPs — except for major commercial ones such as fodder, *tendu* leaves, and certain oil seeds — have not been monitored by the national government. While NTFPs contribute to the livelihood and welfare of millions of people, there is very little research available on their economic value in meeting subsistence needs as nutritional supplements, scarcity foods, housing materials, tools, fertilizer, and medicines.

Ecologists and botanists who have attempted to value subsistence forest-product flows have experienced many problems.[1] The lack of adequate valuation methodologies, which examine the direct and indirect costs and benefits from different forest management systems, makes attempts by the government of India and the donor agency to assess the viability of investment in forest development a flawed exercise. Until better data and more integrated valuation techniques become available, conventional financial analysis provides a faulty basis for forest planning and decision-making. Some scientists have recently attempted to develop more systematic methods of estimating the opportunity costs of NTFP collection to better evaluate forest production activities.[2]

The use of participatory rural appraisal (PRA) and diagnostic research methods can help researchers and foresters learn more from forest communities about forest values, uses, ecology, economics, and local institutions. While often ignored in the past, India's rural women and tribal communities are proving to be a valuable source of indigenous information. Due to their heavy dependence on forest resources, forest users often possess extensive ethnobotanical and ecological wisdom concerning local ecosystems and their products. This essay reviews some preliminary findings from this interactive field-learning and underscores the need for new, locally informed perspectives on the forest and its management.

The Commercial Perspective

The modern science of forest management emerged in Europe in the mid-nineteenth century as a response to the growing shortage of timber needed for construction and fuel purposes. As a consequence, the establishment of an institutional structure in India to manage forest resources was dominated by the colonial objective of enhancing the availability of logs. Initially this involved logging natural forests where logistically possible, and later manipulating the forest to improve timber yields.

Economic evaluations of forests in the past have focused on pricing the standing stock of commercial timber. The 1952 Forest Policy identified national forests as the 'basis of India's strength and wealth', while also stating that the rights of community users must be restricted towards this objective. In 1961, the forest sector contributed 4.4 per cent of the nation's gross domestic product (GDP); by 1989, it had declined to 3.7 per cent. Interestingly, the balance between wood and non-wood products has shifted over the decades: by 1989 only 1.6 per cent was derived from logs and wood products; the remaining share of 2.1 per cent came from non-wood forest products. However, the GDP is a limited indicator of value. It conveys but a small portion of the forest's actual contribution — only those goods that are tangible, marketable, and documented. Such calculations ignore two important factors: the much broader range of qualitative forest benefits and services that elude market valuation, and the proper costing of environmental externalities that result from forest destruction. It has been estimated that the value of ecological and environmental services proffered by the forest as an integral *system* may be eight to ten times its recorded value of goods.[3]

From 1946 to 1985 the commercial forestry sector's contribution to the national domestic product grew by a significant 1.76 per cent per year, despite a steady decline in forest area and biomass productivity, and the substantial increase in the population of poor forest users competing with the industrial sector for forest resources. By contrast, until the early 1980s, reinvestment in the forestry sector has been strikingly low, reflecting a resource mining rather than management orientation.[4] While substantial funding was being made available for social forestry programmes, this investment was used almost entirely for private and community plantations, with little being spent on natural forests. In short, for decades timber management has been synonymous with forest management, limiting researchers' and planners' thinking regarding policies and methods to enhance the local economic and broader environmental functions of natural forests.

The Indian approach to forest valuation and policy-making is practised in many countries. Government forest departments, under instructions to stimulate industrial use, have sold logging leases to commercial firms at prices far below the real value of the timber. Such undervalued pricing policies have encouraged unsustainable use-levels that result in 'boom and bust' extraction practices.[5] Government forest departments set lease rates and stumpage fees for industrial users without considering the social and environmental costs of logging, or the costs of re-establishing the forest. Ignoring these costs distorts decision-making by management.

Forest management policies that have supported short-term commercial extraction have been fostered by inaccurate or incomplete economic assessments. Policies supportive of rapid forest exploitation for industrial use have been encouraged by strong vested-interest groups within the larger political economy of nations. During the British colonial period in India, private sector concerns with political connections greatly influenced utilization plans and biomass flows. This pattern was perpetuated even after

Independence, evolving to encompass the new electoral system. Certain politicians received campaign contributions from business people in return for directing state forest departments to allocate timber lease rights to them. The outcome was an impenetrable, three-party coalition graphically referred to as the 'Iron Triangle'.[6] As a consequence, understanding the economics of forest management requires not only a resource economist's perspective, but an analysis of the political economy influencing forest policy formulation.

Forest management for profit maximization was equated with managing them for optimal timber production. This not only influenced the rates of extraction, but silvicultural methods as well. Colonial management plans attempted to create even-aged, uniform stands of a narrow spectrum of commercially valuable timber species. The objective was to generate more homogeneous 'high forests', culling out less valuable trees and understorey species. At that time, the impact on the biodiversity and ecological values of the natural forest were scarcely considered. In the late 1970s, state forest development corporations (FDCs) were created to meet the growing industrial demands through increased timber production on forest lands. After clear-felling, the FDCs would convert some of these lands to high-value commercial plantations.

Today, the legacy of industrial timber-oriented management policies is mirrored in the continued reliance on the ten- and twenty-year working-plans. The silvicultural prescriptions outlined in the working-plans indicate how deeply ingrained commercial timber production is in Indian forestry. While these plans did provide a long-term framework for timber management planning, they have become increasingly anachronistic. The operations proposed in the plans often do not reflect ground-level realities. Due to inadequate updating, the plans rarely portray actual forest conditions and tend to inflate standing-timber stock figures. Frequently, operations proposed in the plan are not funded under the national Five-Year Development Plans. Finally, working-plans emphasize timber production strategies for large areas, and consequently are not sensitive to local conditions and needs, and generally exclude other management objectives such as generating NTFPs, improving hydrological function and soil conservation, and protecting biodiversity. The narrow perspective of such plans and the inability of forest agencies to follow them due to their limited staff and fiscal resources argue for the adoption of micro-management plans based on local resource realities and community priorities.

In the face of rapidly expanding national demands, medium and large-scale forest-based industries are turning away from public forest lands for raw materials, towards new sources from private farmers and imports. This reversal is crucial, as it signals a sense of resource limits and new management priorities for degraded public forest lands. As illustrated in the 1988 forest policy, the satisfaction of basic, local community needs for fuel, fodder, food, and pasture requires greater priority as India's rural population expands. Only when the definitive link between rural poverty alleviation and sound natural forest management is made, will national industrial interests be forced to look outside the domain of public forest lands to meet their needs.

The Community Perspective

In contrast to urban-based planners and industries, rural communities have a very different perception of the value of the forest. In their world-view, timber is only one of the many products collected, and often of secondary importance. Forests are not simply trees but a habitat used for hunting animals, gathering food, medicinal products, fibres, fuel, fodder, green manure, gums, and materials for tools, ornaments, and handicrafts. The forest provides a private space for retreat, meditation, and social activities. Certain forest tracts have also functioned as a 'land bank', intermittently converted to agriculture, while others play a critical support role in the larger agricultural ecosystem. Multi-tiered natural forests help to moderate harsh temperatures in the summer, slow erosive winds, and break hard laterite soils, allowing water to recharge aquifers and enhance agricultural productivity.

Tribal communities throughout India have been characterized by their heavy dependence on forests for their survival. The traditional tribal perspectives on the tenurial status of forests have differed radically from the conventional industrial-capitalist concepts borrowed from the West. For tribals and other forest-dwellers, forests have been perceived as unowned, open access, or communal resources. Their ecocentric outlook views human beings as only one component in a much grander biotic system of complex inter-relationships between plants, animals, and physical forces. Some tribal groups view themselves as keepers of the forest heritage, passed down from their ancestors and to be perpetuated through generations for posterity. This concept of intergenerational equity has demanded practices combining utilization with conservation in order to ensure sustainability of the system.

This is not to imply that all traditional forest communities have always managed forest resources intensively and sustainably. In the past, in sparsely settled areas, forests could be successfully used or 'managed' extensively with minimal technical or organizational input. After experiencing disturbances from agriculture, grazing, or fires, forests had time to regenerate naturally. At the same time, due to their proximity to the forest, recurring household needs for forest products, and recognition of the forest's ecological services, rural communities have had strong incentives to control excessive exploitation and to manage their resources with care for the future. The acquisition of rich ethnobotanical knowledge has stemmed from the communities' broad-based dependence on and common socio-religious ties to the forest. Through the ages, indigenous management expertise gradually evolved through local trial-and-error experiments with use-practices.

Historical documentation indicates that the processes of rapid deforestation have usually been initiated and driven by external forces that have destabilized traditional management systems. After losing control over their local resources through a breakdown in tenure and protection mechanisms, communities are thrown into competition with outside groups for forest resources and no longer have incentives to ensure that use-practices are sustainable. This is especially true when local forest communities witness the entry of larger, more powerful parties that immediately supersede local authority and proceed to rapidly deplete the forest. Disempowered forest-dwellers often have no choice but to join in the overexploitation. As documented in the earlier chapters of this book, forest communities have begun to openly express their value for the forests as they impose local controls and initiate protection activities. In some cases they forcefully deny access to loggers and fuelwood collectors. Where forest departments have supported these groups under emerging joint forest management programmes, government and community perspectives may find common ground, leading to more sustainable management.

Macroeconomic Assessments

While 95 per cent of India's forest area is under government ownership, the strong dependence of India's large rural populations on these forest resources is propagating new perspectives on the economic costs and benefits of forest management alternatives. Past management models, which have stressed timber production for industrial uses, are being increasingly questioned as the immense demands of the expanding rural population grow. With fifty-four million tribals heavily reliant on forests, in addition to the 250 to 350 million people partially dependent, the government assumes tremendous fiscal liability and social responsibility by underselling forest resources to the industrial sector.

By the year 2000, India's natural resource base will support about one billion people. The vast majority will continue to exist in a 'biomass-based subsistence economy'.[7] This means that across the nation's entire land area, each hectare, including deserts, mountains, and swamplands, will need to support an average of three people. Since one-third to one-half of the country is currently classified as under-productive or non-productive wasteland, the remaining lands will need to carry a heavy and disproportionate share of the human dependency load. While irrigated and fertilized croplands have generally experienced increases in biomass and food yields, some analysts feel these land resources are reaching the limits of their productivity.[8] At the same time, scientists note that India's forest productivity is far below its natural potential. National average forest yields recorded are only 0.5 to 0.7 cubic metre of timber and fuelwood per hectare per year, yet actual productivity in well-protected forests ranges from over 1.0 cubic metre per hectare in central India up to 3.85 cubic metres per hectare in the Western Ghats. Lal suggests that India could produce more than ninety million cubic metres per year, assuming that its forests are uniformly well protected.[9] Yet, is it realistic to assume that India's sixty-one million hectares of forest land can be adequately protected by a relatively small group of forest guards? Experience in recent decades indicates that the forest department alone cannot perform such a difficult task.

While supply levels appear to be well below biological potential, demands are high and growing. An attempt has been made below to provide some rough assessments of current consumption levels of these critical forest products.

Firewood

India's fuelwood needs are immense. In 1985, an estimated 141 million tons were consumed, representing about 200 kilograms per capita.[10] When extrapolated to the 1991 population, firewood demand increases to 204 million tons, 88 per cent of which is required in rural areas. According to India's 1987 *State of the Forest Report*, the recorded production of firewood from forests was 19 million tons, with an additional 30 million tons obtained from private land.[11] The source of the remaining 155 million tons is not clear, but it is probably unofficially extracted from forests and common lands. Given the current condition of India's forests, sustainable yield levels are considered to be about 32 million cubic metres per year — evidence that the nation may be overexploiting its forest ecosystem by 800 per cent.[12] Further, projections indicate that fuelwood demands may double by the year 2020. While some reduction in per capita demand may occur as electrification and fossil-fuels become more widely available, India's growing population will expand the demand for forest-fuel in the foreseeable future. The question regarding ways to control the 'sustainability gap' assumes greater urgency.

Fodder and Grazing Demands

Fodder and grass for livestock is the dominant biomass requirement of rural India. The bovine population in India has increased 31 per cent over the past thirty-five years from 198 million in 1951 to 259 million in 1987. Similarly, sheep and goat population pressures have expanded over 200 per cent from 47 to 144 million during the same period. Thus, livestock grazing pressures in India's forests are about three times the ecosystem's carrying capacity.[13] While livestock populations have grown significantly since Independence, there is probably only half the required area of common land available for grazing.[14] Privatization of the commons, combined with the conversion of forests and commons to industrial and agricultural uses, has been the main reason for the reduction of grazing lands.

The poorest sections of the society — the marginal farmers and the landless with livestock — are most dependent on these forests and commons to satisfy their grazing needs. Although a single hectare of grazing land may support an average of nearly 5 head of cattle, sheep, or goats, grass productivity from the often-degraded lands is low. Grass productivity in Rajasthan, Gujarat, and Tamil Nadu ranges from 0.4 ton to 1.6 tons per hectare per year. Well-protected grasslands normally produce at least 4 to 5 tons per hectare.

The current combined availability of green fodder from all sources in the country, including agricultural lands, pastures, and forests, is estimated to be 434 million tons, as compared to an estimated minimum demand of 992 million tons in 1990.[15] Many specialists feel that the gap between supply and demand is largely met through overgrazing the nation's forest lands. Livestock pressures have resulted in a vicious cycle of land degradation leading to soil erosion and compaction, declining grass productivity, and extensive grazing pressures. Owing to increases in population, greater commercialization, certain public policies, technological changes, and environmental pressures, land degradation has also been accentuated by the weakening or collapse of many traditional forms of common property resource management systems. A serious commitment to addressing the issues of pasture-land regeneration and productivity, as well as sustainable grazing management in forests, is now imperative, as livestock pressures in India will undoubtedly continue to increase in the coming years.

Non-timber Forest Products

Non-timber forest products (NTFPs) are estimated to generate 70 per cent of all employment in the forestry sector. Commercial NTFPs alone are estimated to generate Rs 3 billion ($100 million) annually.[16] Local communities, as well as industries, depend on forests for a wide range of NTFPs. Most NTFPs are consumed directly by rural households; however, the value of those marketed to industries is immense. The major products are bamboo, non-edible oil seeds, tendu leaves, gums, resins, grasses, fibres, and fruits.

Perhaps more important, the collection, processing, and marketing of NTFPs provide employment for millions of rural Indians, who would otherwise be forced to migrate in search of jobs as agricultural labourers or marginal urban job seekers. One study has estimated that NTFP collection generates more than 2 million person-years of work annually and that this could be increased to 4.5 million person employment years.[17] In addition, millions of individuals are employed in NTFP processing and marketing.

The degradation of forests has seriously affected the diversity of NTFP availability and, in turn, the livelihood of the forest-dependent population. Bamboo and oilseeds, in particular, have immense potential that has not been actualized due to poor forest management practices and the lack of effective policies and programmes supporting processing and marketing. For example, major oilseed species like mahua, *neem, karanj, kusum,* and sal generated only 0.5 million ton of oil in 1981–82; yet, the Forest Research Institute calculates that through better management, production could be increased to nearly 6.8 million tons. This would not only create several million jobs but reduce the current import of essential oils.[18]

Despite the importance of NTFPs to both the national economy and to tribal and forest community livelihoods, enhancing their production and marketing has been given scant attention until recently. In fact, the production potential of NTFPs has been estimated at 2.5 to 10 times the current levels.[19] Still, except for a few major NTFPs such as tendu leaves, research into the silvicultural requirements for enhancing productivity and market opportunities has been woefully limited. If guided and technically advised by the government on price information, markets, and infrastructure, the scope for NTFP improvements through value additions and better marketing is vast. Higher returns from NTFPs can not only directly increase household incomes but also generate employment, improve women's welfare and labour burdens, and ameliorate the environment by taking some of the pressures from overgrazing and fuelwood headloading off pastures and forests. In short, NTFPs play a major role in the formal forest sector economy, while employing millions of rural people in the informal sector. Developing management objectives and systems, which emphasize NTFP production over timber, would seem a logical response to the need to increase rural employment opportunities and stabilize natural forest resources.

Timber

India's annual demand for structural timber between 1981 and 1985 was variously estimated at between 19.5 and 30 million cubic metres.[20] Industrial demands for softwood for paper, rayon, and packaging are expected to more than double in twenty-five years from 12 million tons to 25 million tons per year. Currently India depends on imports to meet its forest product needs. Between 1986 and 1989, the cost of imported timber increased nearly tenfold to Rs 3.5 billion ($113 million).[21] Softwood demand could be met from clear-felling of forests like bamboo or tree plantations like eucalyptus. Current estimates indicate that even if a small percentage of private farmlands near industrial areas were planted with fast-growing trees, they could supply the nation's softwood demand, reducing pressures on natural forests. It is less clear how hardwood demands might be met. Nonetheless, farm forestry has immense potential in India, and its expansion in the 1980s was a healthy response to the growing market demands. The emphasis on timber production from farmlands, while offering opportunities to reduce pressures on natural forests, does not provide answers to how the requirement for fuelwood might be met. India remains the world's largest consumer of wood, largely due to its immense demand for fuelwood. Combined with wood extraction for industrial timber and other uses, some estimates indicate that more than 264 million cubic meters of wood was harvested in 1988.[22] In 1986, more than 87 per cent of all wood used nationwide was consumed as fuel.

While India's natural forests have suffered under the increasing demands for commercial timber, the nation has already made great strides in privatizing industrial forestry. Between 1981 and 1988, Indian farmers planted 8.6 billion trees on private lands. With a survival rate of 60 per cent and a planting density of 2,000 trees per hectare, an estimated 2.5 million hectares of farmland is already under commercial timber. Farmers surprised government planners and donor agencies by far exceeding planting targets in response to the attractive pole and timber market prices. 'In the north-west region, which has a long history of growing agricultural crops for markets, as much as 4 to 6 per cent of cultivable land was brought under trees in a short spell of 5 years.'[23] The private tree plantings of the 1980s should yield about 10 million tons. Unfortunately, market fluctuations, felling and transit regulations, and inadequate information and extension services have frustrated many private tree farmers. Farmers are reluctant to wait long enough for the trees to mature to the commercial size of more than 20 centimetres in diameter. In Haryana, only 7 per cent of the 1989–90 tree harvest could be classified as timber. Farmers also face constraints in selling their trees for pulpwood, firewood, and pole markets. Nonetheless, farm forestry in India has expanded rapidly in the 1980s and is likely to grow and develop in the future. Supportive policies and programmes can facilitate this development, reducing commercial pressures on natural forests, benefiting private farmers and forest dwellers.

Soil Conservation

A critical environmental externality, which remains unreflected in equations of state profits from forest exploitation, is the enormous loss of India's productive topsoil. Efforts to reverse the massive soil erosion have become a resource management imperative for this predominantly agrarian nation. An estimated 100 million hectares of productive agricultural land experience qualitative decline in soil fertility each year. Topsoil losses in India are estimated at between 16 and 20 tons per hectare, with national losses estimated at between 5300 and 12,000 million tons annually.[24] Severe soil erosion eventually renders arable land completely unproductive. This phenomenon is being witnessed in an extreme form in the states of Uttar Pradesh, Madhya Pradesh, Bihar, Rajasthan, and Gujarat where ravines are increasing by 8000 hectares every year, adding to the more than 4 million hectares of existing ravine-and-gullied territory across the country. Eliminating arable lands from food production has particularly dire consequences for a country yet to come up with solutions to check its population growth rate. Despite more than a 50 per cent increase in the total crop yields from the 1960s to the mid-1980s, the human population increased by more than 72 per cent. This exponential population increase has essentially outstripped agricultural gains leading to a reduction in the per capita food consumption.[25]

In addition to the loss of productivity — which India can ill afford — massive soil erosion results in downstream sedimentation loads in hydroelectric and irrigation reservoirs, adversely affecting their productivity. Constructed in 1973, the anticipated life span of the Ram Ganga dam in Madhya Pradesh has now been reduced from 185 years to forty-eight years due to heavy siltation in its feeder rivers. It is estimated that nearly 20 per cent of India's water-storage capacity will be eliminated by the year 2000, with a corresponding loss of 4 million hectares of irrigable land.[26]

The important water-absorption function of a robust natural forest has been well documented. Hydrological regulation is particularly significant in India, which is highly subject to periodic flooding of vulnerable lands. India's flood-prone area has expanded from 20 to 40 million hectares during the past two decades. It is not the trees *per se*, but the ground cover and multi-tiered understorey structure of the natural forest that play a key role in stabilizing the soil and controlling surface run-off. Alternative land uses, such as intensively managed agroforestry tracts with vegetative bunding, can provide excellent soil and water conservation in addition to high economic benefits; in lieu of tenure security and the widespread transfer of such intensive management technologies to farmers, the natural forest offers the highest level of soil and water conservation services for a given level of land management. Considering that nearly 68 per cent of India's agricultural land, or an estimated 96 million hectares, has no irrigation facilities, maintaining a good forest cover to reduce soil erosion, slow water run-off, and generate mulching materials is of great importance.[27]

Biodiversity and Environmental Stability

India's natural forests also harbour the nation's impressive biodiversity. With more than 65,000 species of fauna, 45,000 species of flora, and a high degree of endemism, the escalating human and livestock pressures on India's forests have made these habitats among the most threatened in the world today. The protection of biological diversity, most of which lies in the natural forests, serves a higher purpose, that of global carbon sequestration. The standing forest serves as a carbon 'sink' — until it is burned and becomes another carbon source — and deforestation may cause up to one-fourth of all global carbon emissions into the atmosphere. In the Amazon forests, for example, 220 tons of carbon is released when the slash from each hectare of forest is burned.[28] Preliminary research is now suggesting that the degraded but regenerating, multi-storied secondary forest may actually sequester more carbon than an older, undisturbed forest tract — a finding with important implications for promoting natural forest regeneration under community protection schemes. Certainly, the carbon sequestering capacity of forests far exceeds that of grasslands. Recent studies indicate that 'pastures hold only 3 per cent of the above-ground carbon and 20 per cent of the root carbon that forests do.'[29]

In contrast to a clear-felled or otherwise degrading forest, one that is sustainably managed to generate a regular flow of forest products can offer consistent rural employment opportunities. This is a significant alternative in the light of India's current economic trends that indicate a serious decline in employment growth in the rural sector, particularly in agriculture and for women. Because the employment content of rural economic growth is declining, the idea that economic growth will take the pressure off the forest resource base is unrealistic, at least in the near future.

National Security and Social Stability

Resource-based conflicts can have high economic costs. Conflicts over forest access affect the daily lives of thousands and have done so for decades. Disputes over use rights cause foresters to patrol more intensively, plant species of less value, install costly fences, and become involved in fights that may injure them, resulting in loss of work time. Villagers have to walk additional distances to avoid contact with forest guards, sell their forest products at discounted prices because they are illegally gathered, and face fines and imprisonment if apprehended. N.C. Saxena reports the story of Dapubai, a tribal woman from Udaipur who received only Rs 7 for 10 kilograms of gum that took her ten days to collect. When asked why she had not received the market price of Rs 250, she responded, 'How can I demand a higher price? The trader's man threatens to report me to the Forest authorities for entering the forest . . . Then we will get nothing.'[30]

Conflicts have a significantly adverse impact on the time allocation and resource decision-making of all individuals involved. Rather than planting a tree for future benefits, individuals often realistically assume that they should overexploit the forest in the present to ensure that they gain some benefit, before others deplete the resource. Conflict can also lead to institutional inefficiency. For example, in Rajasthan, the number of cases filed in court for illicit felling and grazing increased from 18,600 cases in 1961 to 35,788 cases in 1981.[31] Nationwide, thousands of cases are lodged annually, clogging India's judicial system and making it less effective. In situations where conflicts over resources become an established part of the social environment, they may become a focal point for armed separatist and insurgency movements.

How can the value of a nation's functioning forests be evaluated in terms of the types of economic, social, and environmental services described above? J.B. Lal has attempted this difficult task and estimates that the annual value of forest goods and services, including timber, firewood, NTFPs, grazing, biodiversity, recreation and environmental services, would total Rs 800 billion ($27 billion). This is more than one-fourth of India's 1987–88 gross domestic product, far higher than forestry's official contribution of 1.2 per cent.[32] While economists will continue to debate forest valuation procedures and statistics, few would argue that a sustainably managed natural forest environment is critical to the nation's social, political, and economic future.

Microlevel Economic Assessments: Community and Forest

The preceding macro-perspective on the value of India's forests outlines the vast scale and numeric importance of the resource to the nation. For the community, neighbourhood, and households forests take on a different significance and complexity. Forest products flow to millions of rural Indians but are particularly pertinent to poorer households, especially for women. The role of the forest in their daily lives changes by season and year. In the dry season, when croplands wither under the sun, the forest provides food and fodder. During drought and famine years, forest resources become critically important. Due to the many roles that the forest plays in rural life, it will be helpful to examine its significance in terms of the kitchen and home, the farm, the market, and the social community.

Forest and the Home

In remote Indian communities, where cash is scarce and markets distant, forests function much as shops do in urban areas. The forest meets diverse rural household needs for food, fuelwood, medicines, housing, tools, toys, utensils, and other goods. One study from West Bengal indicated that even in seriously degraded forests, after a few years of regeneration, communities were using 72 per cent of the 214 species present. Of the 109 different items routinely used by village households, only a small number of goods were purchased in local markets. Researchers have estimated the value of the average household's consumption of NTFPs at Rs 2500 to 2700 ($80–87) per year, representing approximately 20 per cent of all family income.[33] However, the real value may be far greater, since forest products are available in times of economic stress, providing famine foods, fuel, and medicines when no cash is available to purchase these goods from other sources.

A study from Tamil Nadu in southern India produced similar findings. Villagers living near Kadavakurichi forest used virtually all of the sixty-three species present, including one-third for fuel, 17 per cent for fodder, 40 per cent for medicinal purposes, 30 per cent for food, and 19 per cent for agricultural tools and household articles. The study found that the forest generated NTFPs valued at about Rs 2100 per hectare per year.[34]

Although foods from the forest are often viewed by urban-based researchers and development specialists as 'famine foods,' they can be nutritious, flavourful, and a major source of sustenance through much of the year for millions of forest families. In states with extensive forest resources like Orissa, Madhya Pradesh, Himachal Pradesh, and Bihar, up to 80 per cent of forest dwellers depend on the forest for 25–50 per cent of their food needs.[35] The Saura tribal community in Mahapada village in Dhenkanal District of Orissa, for example, depends on forest tubers (*Dioscoria* spp.), mushrooms, edible leaves, and fruits for six months of the year.[36]

In central and western India the mahua (*Madhuca latifolia*) is revered, due to its importance as a source of food, fodder, oil, and fuel. The flowers are eaten fresh, distilled for liquor, and collected, dried, and made into powder for preparing unleavened bread. Bamboo shoots (*Dendrocalamus* spp.), palm stalks (*Corypha umbraculifera*), palm sap (*Elate sylvestris*), and jack fruit (*Artocarpus heterophyllus*) are important dietary supplements in rural India.

The importance of forest foods for villagers varies widely depending on the condition and diversity of the ecosystem and a group's traditions and preferences. In India, tribal communities are often the most sophisticated connoisseurs of forest foods. Forest fruits provide critically needed vitamins and minerals, often in high concentrations. *Zizyphus jujuba*, a common snack for village children, contains nearly twenty times the vitamin C available from the same quantity of oranges.[37] Many insect larvae are extremely high in protein. Forest leaves are often high in vitamins A and C, as well as calcium, niacin, and iron. Sal, tendu (*Diospyrous melanoxylon*), and mahua seeds are often pressed and the oil extracted is used for cooking purposes. Sal seeds, which have a 12.5 per cent oil content of high quality, are being developed for export for use in chocolate manufacturing.[38]

In India and in much of the developing world, 'forests often provide the only medicines available to the vast majority' of the population.[39] While their efficacy varies, many plants contain high concentrations of the chemicals that form the base of modern drugs. In addition, they are frequently high in vitamins and minerals, which counteract diseases originating from nutritional deficiencies. Forest dwellers often have a highly developed knowledge of the properties of medicinal plants. One study in southwest Bengal indicated that of the 200 species growing in the neighbouring regenerating sal forest, forty-six were commonly used for medicinal purposes. Local medical practitioners were not only able to identify the species, but also the part of the plant most effective in treating specific illnesses, whether it be the leaf, flower, fruit, seed, twig, pod, stem, root, or bark.[40]

Forests also supply the raw materials needed in the construction of rural houses throughout India. Timber and poles are used for making frames for homes. Where timber is available, shingles are split for roofing. In other areas, large leaves or forest grasses are used for thatch roofing. Villagers in south-western Bengal reported experiencing severe housing problems when the deforestation of neighbouring forests reduced the availability of roofing poles, which need periodic replacement due to termite damage. They noted that their decision to impose strict forest access controls was motivated in part by their need to ensure a stable supply of construction materials.

One of the most critical resources supplied by the forest is energy for lighting, cooking, and small industry. With the increasing disappearance of commercial timber, more than 90 per cent of the wood extracted from India's forests is used as fuelwood.[41] Less than one-sixth of this is legally gathered from any type of managed or monitored supply. In rural areas across India, fuelwood contributes 66 per cent to household energy-use among all income groups, and a much higher percentage to the most disadvantaged — the poorest of whom use mostly twigs and leaves for cooking. Only 23 per cent fuelwood in rural areas is purchased, compared to 75 per cent in urban areas. The remaining huge quantities of woody biomass are collected *gratis*, predominantly from state forest and revenue lands by an estimated 2 to 3 million women and children, an activity that takes up to five hours every day.

Forest and the Farm

The forest supports the larger agro-ecosystem in numerous ways. Particularly, in the mountainous and hilly regions such as the Himalayas and Western Ghats, where steep slopes under cultivation are more prone to erosion and fertility-loss, nutrient transfers from forests to fields allow agriculture to be practised on a sustainable basis. Green manure, particularly leaves, are collected from forests, composted, and applied to the farmers' fields, providing both nutrients and organic material to enrich the soil's chemistry and structure. In the Western Ghats, one study found that 7.6 metric tons of biomass were harvested per hectare from *Soppina betta* forests, of which about 6.5 metric tons was green and dry leaves and grass composted and used to enrich areca-nut gardens.[42] For many marginal farmers, when commercial fertilizers are unaffordable, the use of green manure recycles essential nutrients back into fields, enhancing crop productivity while conserving soils and water moisture. Yet, if pressures become too great, stabilizing agriculture by transferring nutrients can draw down forest ecosystems. In the Western Ghats, although forest biomass extraction is high, studies indicate that it may be sustainable if farmers invest sufficient time in promoting forest regeneration through protection and enrichment planting.[43]

Forests also provide the land resources for shifting cultivation (jhum), a form of agriculture practiced by millions of people in eastern and north-eastern India. Often stigmatized by foresters as a major force behind ecological degradation, swidden farming has been practised sustainably for centuries. In recent years, however, due to growing population and land use pressures, many shifting cultivators have been forced to reduce their fallow periods from longer cycles of ten to twenty years to a shorter rotation of five years or less. Despite past criticisms, recent studies indicate that swidden farming is energy-efficient with positive scientific features, offering 'much scope for redevelopment to correct distortions brought about by shortened agricultural cycles', both in terms of economic yields and land degradation.[44] It is recommended that nitrogen loss during the planting period be recovered through the introduction of nitrogen-fixing trees once the cropping period has ended — a form of assisted or accelerated natural regeneration.

Other direct contributions from the forest to the farmer are well known. One is the regulation of hydrological and moisture balance, including the slow absorption and release of groundwater to crops, flood and drought control, and influence over the moisture levels of local and regional micro-climates. The presence of higher levels of soil moisture for even a few weeks can mean the difference between one and two crops per year in many rainfed areas. By facilitating water infiltration, slowing winds that accelerate evaporation, and providing greater shade, forests adjacent to farmlands can play an important role in improving moisture conditions and limiting wind erosion. A frequent indirect benefit from the surrounding natural forest is its role in pest control. By supporting habitat for birds and other useful predator-prey relationships, forests tend to reduce damage to crops by pest populations.

Forests are also important sources of wood for agricultural implements such as ploughs, hoe and axe handles, storage containers, bullock carts, and waterwheels. Forest communities carefully select different forest species for agricultural implements depending on their hardness, grain, and flexibility. In India, axlewood (*Anogeissus latifolia* and *A. pendula*) is highly desirable for ploughs, water-wheels, and housing beams, whereas *Acacia leucophloae, A. nilotica,* and *A. catechu* are commonly used for hoes and axe handles.

Finally, the forest serves as a vast reservoir for grazing and fodder collection in India. Poor landless families graze their cattle on common property, including forest lands.[45] Communal lands have decreased steadily in recent decades, placing greater pressure on the remaining forest lands. In western Rajasthan, for example, grazing lands decreased by 16 per cent between the mid-1950s and mid-1970s, while the livestock population increased by 40 per cent (by 6 million). With more than half of the world's buffaloes, 15 per cent of the earth's cattle and goats on one-fortieth of the globe's land, pressures on biomass resources for fodder are intense.[46] Forest lands have been critically important in supporting India's immense livestock populations, but they have also suffered in the process. Community-based controls and regulations on grazing in protected forest lands in eastern India have been effective in stabilizing pasturage and presenting new opportunities to regain productivity.

Forest and Village Industry

Estimates of labour absorption in the forestry sector are variable, but there is a general agreement that forest employment is extremely important to millions of low-income people. In tribal areas the dependency on NTFPs is even higher: 'Thirty-five per cent of the earnings of tribals in a district of Gujarat and up to 38 per cent in Madhya Pradesh have been found to come from these items.'[47]

One study reported that bamboo and cane collection and processing provide about 400 million person-days of work annually, whereas the collection of non-timber products generates 280 million person-days, and fuelwood and charcoal collection, processing, and marketing an additional 200 million person-days.[48] An estimated 15 million people are involved full time in the commercial fuelwood trade generating up to 4.5 billion person-days of employment per year.[49] If fuelwood headloading were to cease due to depletion of forest resources, finding alternative jobs for those displaced would be difficult.

Tendu leaf collection is a major source of employment for tribal people living near natural forests throughout much of central India. Its leaves are used as wrappers for *bidis.* An estimated 7.5 million people are involved in collecting leaves during the three-month harvest season. In addition, another 3 million people are employed throughout the year in the bidi processing industry.[50]

Many forest products are also critically important as raw materials for cottage industries. For example, there are probably well over 100 million string cots (*charpoi*) in India, most of which are woven with handmade grass rope (*ban*). They are generally restrung every 2–3 years with 2–3 kilograms of rope. Although market surveys have not yet been conducted, it is likely that between 50 and 100 million kilograms of grass rope manufactured from forest grasses (bhabbar or *sabai*) is made in India each year. This activity employs millions of rural men and women throughout the country, yet there is little recognition of this employment sector in government economic records. Similarly, bamboo and reed mat and basket-making employ millions in rural India. In past decades, Indian forestry has faced difficulties in ensuring a steady supply of reasonably priced raw materials for artisans. As a result, those employed in forest-based cottage industries have had to depend on illicit or low-quality materials. This situation has also encouraged poor management, over-exploitation, and sometimes threatened local industries with collapse due to the erratic supplies of raw materials.

Research indicates that the poor in semi-arid regions depend on common property resources for up to 84 per cent of their income.[51] Other studies suggest that NTFPs alone can generate up to 55 per cent of the total income of a tribal family. This is the case even though many cottage industries based on NTFPs are commonly limited by raw material supplies, monopolies by middlemen, and poor market information and infrastructure. There is a vast potential for improving NTFP productivity, equity to collectors, and market efficiency.

With millions employed in the non-timber forestry sector, increased efficiency in processing and marketing can result in greater profitability to producers. Just as productivity can be greatly improved through intensive management, so can processing and marketing through better technologies, design, market research, and open market policies. Unfortunately, despite the fact that NTFPs accounted for 41 per cent of all forest revenues and 71 per cent of all forest exports in 1975, when national statistics were last available, investment in research and extension has been minimal.[52] Many NTFPs are sold in their raw form. Since most profits from these products come through processing, forest communities tend to earn very low levels of income from forest-based activities. This need not be the case, however, since processing of many products can be performed by the communities. Simple oilseed presses, distilling equipment, pulp and paper processing technologies, sawn log pit mills, drying racks, and other appropriate tools have already been developed for such applications.

Where traditional processing technologies exist, technical improvements also provide considerable scope for increased efficiency. Inexpensive pedal-driven and electrically driven rope-weaving machines can improve output by 500 to 1000 per cent.[53] In West Bengal where thousands of women obtain a substantial part of their income from stitching sal leaf plates, presses, which thermally weld plastic inserts between two layers of leaves while shaping the leaves into a bowl, have been introduced. This process produces a drip-resistant container, which is far better received in markets for weddings and ceremonial gatherings. The machine-processed plates with inserts are sold in Calcutta at ten times the price (Rs 20–24 per 100) that traditional plates bring.[54]

The efficient marketing of timber and NTFPs has suffered from poor policies and inappropriate regulations. In a sincere attempt to control misbehaviour by contractors and middlemen, the government established a variety of trade and transit regulations. Many states also established government corporations to oversee trade in important NTFPs. Often, the result has been the imposition of bureaucratic controls that have constricted free trade. Village collectors frequently find themselves at the mercy of a single, state-licensed buyer who offers uncompetitive prices for their products. For example, the tendu leaf trade is highly controlled in virtually all major producer states. State governments have passed laws requiring registration of all growers, traders, and bidi manufacturers. Agents are appointed by the states and given exclusive rights to purchase leaves, wage rates for collection are fixed by committees of buyers and state agents, while transport permits are required for any movement either within or outside the state.[55] Collusion among state agents, purchasers, and manufacturers is common, while wage rates are often fixed at levels far below the minimum agricultural wage.[56] In essence, regulations are manipulated to serve the needs of the more powerful participants, while the poor village collectors have generally continued to be exploited.

Many argue that forest communities would be better off with an open market system where buyers would compete for their products.[57] Simple deregulation may be the most effective way to bring better NTFP prices to forest communities. In addition, providing guidance to collector groups on the methods of grading, drying, and storage would put them in a better position to market their own products more strategically. Trade agencies, which provide information on market prices, seasonal fluctuations, and purchasers, might better empower collectors than forcing them to sell to a single buyer.

Community Managed Forests and New Economic Perspectives

Given the growth of India's population and its continuing reliance on natural biomass, the productivity of the nation's natural forests will need to be optimized through more intensive management. Currently, much of the country's forests generate only a fraction of their potential, whether it be timber, pulp, grass, fruits, oilseeds, or other valuable products. With the country's population approaching 1 billion — and continuing to grow — and given the finite availability of soil, water, and forest resources, these essential elements will need to be managed with increasing efficiency. Besides increasing biological productivity, employment and income-generation from forest-based activities can also be enhanced through improved processing and marketing.

As India enters a new millennium, the nation is presented with opportunities to adjust its forest management policies to respond more effectively to changing social, economic, and ecological needs. Where current use practices are leading to forest degradation, there is a special need for management reform. One option would be a gradual shift of all commercial wood production to privately owned lands. India's impressive success with farm forestry indicates that private timber producers are far more effective than public enterprises. In Karnataka, government eucalyptus plantations frequently yield only 1.5 to 2 metric tons per hectare, as against 10 to 15 metric tons per hectare from private lands. It has been estimated that only 3 to 6 million hectares, comprising less than 4 per cent of all cultivated land, could meet the 30 million ton timber requirement of the commercial sector.[58] By producing most commercial timber on private lands, logging pressures on India's 60 million hectares of natural forests could be greatly reduced, allowing these ecosystems to recover. Natural forests, freed from the burden of supplying the commercial timber sector, could be sustainably and intensively managed by communities, meeting local forest product needs, supplying raw materials for cottage industries, and stabilizing local agro-ecological and hydrological systems. The value of these services far outweighs revenues generated from timber. If the nation's natural forests are managed more intensively, forest contributions to rural employment, agriculture, water conservation, and other sectors can be dramatically increased.

Notes

1. C.M. Peter, A.H. Gentry, and R. Mendelson, 'Valuation of a Tropical Forest in Peruvian Amazonia', *Nature*, 339:655–6, 1989.
2. R. Godoy, R. Lubowski, and A. Markandya, 'A Method for the Economic Valuation on Non-timber Tropical Forest Products', *Economic Botany*, 47:220–33, 1993.
3. Steven Schonberger, 'Linking the Environmental, Industrial, and Local Benefits from India's Forests', unpublished manuscript, 1992.
4. R.A. Sharma, 'Man's Impact on the Forests of India', *Wasteland News* 8(3): 11, August–October 1992.
5. Jeffrey R. Vincent, 'The Tropical Timber Trade and Sustainable Development', *Science*, vol. 256, 19 June 1992.
6. Madhav Gadgil, 'Deforestation: Problems and Prospects', Founder's Day Lecture, Society for the Promotion of Wastelands Development (New Delhi, 12 May 1989).
7. Anil Agarwal and Sunita Narain, *Towards Green Villages* (New Delhi: Centre for Science and Environment, 1989), p. 1.
8. Ibid., p. 2.
9. J.B. Lal, 'Economic Value of India's Forest Stock', in Anil Agarwal (ed.),

The Price of Forests (New Delhi: Centre for Science and Environment, 1992), p. 45.

10. Tirath Gupta and Vinod Ahuja, 'Demand Forecasts of Household Energy and Policy for Wasteland Management', in Anil Agarwal (ed.), *The Price of Forests* (New Delhi: Centre for Science and Environment, 1992), p. 90.

11. Government of India, *The State of Forest Report* (Dehra Dun: Forest Survey of India, Ministry of Environment and Forests, 1987).

12. See Steven Schonberger, 'Linking the Environmental, Industrial, and Local Benefits from India's Forests', p. 12.

13. See J.B. Lal, 'Economic Value of India's Forest Stock', in A. Agarwal (ed.), *The Price of Forests*, p. 44.

14. J.S. Jodha, 'Common Property Resources and Rural Poor in Dry Regions of India', *Economic and Political Weekly*, 21, pp. 1160–81, 1986.

15. Panjab Singh, 'Forage Research: Present Status and Future Strategy', in *Pasture and Forage Crops Research* (New Delhi: Third International Rangeland Congress, 1988), p. 1.

16. M.P. Shiva, 'Minor Forest Products in the Tribal Economy: Programmes, Problems, and Prospects', in Anil Agarwal (ed.), *The Price of Forests* (New Delhi: Centre for Science and Environment, 1992), p. 113.

17. Tirath Gupta and Amar Guleria, *Non-wood Forest Products in India* (New Delhi: Oxford and IBH Publishing, 1982), p. 133.

18. Robert Chambers, N.C. Saxena, and Tushaar Shah, *To the Hands of the Poor: Water and Trees* (New Delhi: Oxford and IBH Publishing, 1989), pp. 49–50.

19. See Tirath Gupta and Amar Guleria, *Non-wood Forest Products in India.*

20. R.V. Singh, 'Timber Demand in India: Prospects for Future Supply and Substitution', in Anil Agarwal (ed.), *The Price of Forests* (New Delhi: Centre for Science and Environment, 1992), p. 66.

21. Ashbindu Singh, 'Trends in Pulp and Timber Import', in Anil Agarwal (ed.), *The Price of Forests* (New Delhi: Centre for Science and Environment, 1992), p. 72.

22. World Bank, *India Forest Sector Review* (Washington, D.C.: The World Bank, 1993).

23. N.C. Saxena, 'Marketing Constraints for Eucalyptus from Farm Lands in India', *Agroforestry Systems*, 13:73–85, 1991.

24. Anil Agarwal, 'Between Need and Greed: The Wasting of India; The Greening of India', in Anil Agarwal, Darryl D'Monte, and Ujwala Samarth (eds), *The Fight for Survival* (New Delhi: Centre for Science and Environment, 1987), p. 172. Also see B.B. Vohra, 'The Management of Natural Resources', INTACH Environmental Series no. 4 (New Delhi, 1987), p. 6.

25. See Steven Schonberger, 'Linking the Environmental, Industrial, and Local Benefits from India's Forests'.

26. Kunwar Jalees, 'Loss of Productive Soil in India', *International Journal of Environmental Studies*, vol. 24, pp. 245–50, 1985.

27. B.B. Vohra, 'The Management of Natural Resources', INTACH Environmental Series no. 4 (New Delhi, 1987), p. 4.

28. Marguerite Holloway, 'Sustaining the Amazon', *Scientific America,* July 1993, p. 96.

29. Ibid.

30. See Robert Chambers, N.C. Saxena, and Tushaar Shah, *To the Hands of the Poor,* p. 148.

31. Jagdeesh C. Kalla, 'Forest Land Management: A Case Study of Rajasthan', in Anil Agarwal (ed.), *The Price of Forests* (New Delhi: Centre for Science and Environment, 1992), p. 159.

32. See J.B. Lal, 'Economic Value of India's Forest Stock', in *The Price of Forests,* p. 47.

33. Society for the Promotion of Wastelands Development, *Joint Forest Management: Concept and Opportunity* (New Delhi: SPWD, 1992), p. 11.

34. Paul P. Appasamy, 'Role of Non-timber Forest Products in a Subsistence Economy: The Case of a Joint Forestry Project in India', *Economic Botany,* 47(3): 263–4, 1993.

35. Centre for Science and Environment, *The State of India's Environment 1984–1985* (New Delhi: CSE), p. 393, as cited in Jeffrey Y. Campbell, 'Putting People's Products First: Multiple Use Management for Non-wood Forest Products in India', unpublished manuscript, June 1989.

36. Mark Poffenberger, 'The Resurgence of Community Forest Management: Case Studies from Eastern India' (prepared for the Liz Claiborne and Art Ortenberg Foundation Community-Based Conservation Workshop, Airlie, Virginia, 18–22 October 1993), p. 14.

37. FAO, 'Household Food Security and Forestry: An Analysis of Socio-Economic Issues', *Forest, Trees, and People* (Rome, 1989), p. 12.

38. Jeffrey Y. Campbell, 'Putting People's Products First: Multiple Use Management for Non-wood Forest Products in India', draft manuscript, June 1989.

39. See FAO, 'Household Food Security and Forestry', in *Forest, Trees, and People,* p. 29.

40. K.C. Malhotra, 'Role of Non-timber Forest Produce in a Village Economy: A Household Survey in Jamboni Range, Midnapore District, West Bengal', IBRAD Working Paper (Calcutta, 1992).

41. See Steven Schonberger, 'Linking the Environment'.

42. M.V. Nadkarni, *The Political Economy of Forest Use and Management* (New Delhi: Sage Publications, 1989), p. 125.

43. M.V. Nadkarni and S.A. Pasha, 'Contribution of Forests to Agriculture: A Study of Arecanut Gardens in Uttara Kannada', in Anil Agarwal (ed.), *The Price of Forests* (New Delhi: Centre for Science and Environment, 1992), p. 300.

44. P.S. Ramakrishnan, 'Jhum: Is There a Way Out?' in Anil Agarwal (ed.), *The Price of Forests* (New Delhi: Centre for Science and Environment, 1992), p. 305.

45. Hooly Brough, 'Holy Cows, Unholy Trouble', *Worldwatch*, 5(5): 19–21, September 1991.

46. Centre for Science and the Environment, *The Wrath of Nature* (New Delhi: CSE, 1987).

47. Government of India, 'Report of Committee on Forest and Tribals in India' (New Delhi: Ministry of Home Affairs, Government of India), cited in Robert Chambers et al., *To the Hands of the Poor*, p. 144.

48. Arvind Khare, 'Small-Scale Forest Enterprises in India' (New Delhi: Society for the Promotion of Wastelands Development, 1987).

49. A. Van Buren and Peter May, 'Rural Energy Concepts: The Commercialization of Wood and Other Subsistence Fuels' (Brussels: ISPRA, 1982), as cited in 'Household Food Security and Forestry' in *Forest, Trees, and People*, p. 57.

50. D.N. Tewari, 'Minor Forest Products of India' (paper presented at the USAID Conference on Forestry and Development in Asia, Bangalore, April 1982), as cited in 'Household Food Security and Forestry', p. 56.

51. See J.S. Jodha, 'Common Property Resources', *Economic and Political Weekly*, 21.

52. Rashmi Pachauri, 'Sal Plate Processing and Marketing in West Bengal', Working Paper Series no. 12 (New Delhi: Ford Foundation, 1991), p. 2.

53. Mark Poffenberger and Madhu Sarin, 'Fibre Grass from Forest Land: A Case from North India'. Working Paper Series no. 10 (New Delhi: Ford Foundation, 1990).

54. See Rashmi Pachauri, 'Sal Plate Processing and Marketing in West Bengal', pp. 12–13.

55. Tirath Gupta and Amar Guleria, *Some Economic and Management Aspects of a Non-wood Forest Product in India: Tendu Leaves* (New Delhi: Oxford and IBH Publishing, 1982), p. 94.

56. Ibid., pp. 110–14.

57. M.D. Mistry, 'The Impact of the Forest Act on the Household Economy of the Tribals', in Anil Agarwal (ed.), *The Price of Forests* (New Delhi: Centre for Science and Environment, 1992), p. 122.

58. Madhav Gadgil, 'Forestry Principle: What Should the Indian Response Be?' *Wasteland News*, 8(2): 14, November 1992–January 1993.

Chapter 10

Ecological Stabilization and Community Needs: Managing India's Forest by Objective

N.H. Ravindranath ■ Madhav Gadgil ■ Jeff Campbell

This chapter examines patterns of forest degradation prevalent in India. It also explores the potential to restore forest ecosystems if disturbances can be reduced or stopped, and ways in which natural forests might be manipulated to provide a sustainable and increased flow of products to communities. The first step to sustainable forest management will involve halting the degradation process. Once access controls imposed by communities are established, in many areas rapid regeneration will be initiated as a second step to ecological recovery. The third step in the process will require human ingenuity to develop site-specific methods to manipulate natural ecosystems to ensure a sustainable supply of important forest products while enhancing biodiversity, water, and soil conservation. This will often require a reversal of past and current forest management systems.

India's forests have historically been endowed with vast and varied flora and fauna. India is comprised of ten broad biogeographic zones, possessing approximately 15,000 flowering plants, and nearly 1200 species of birds and 341 of mammals.[1] The rapid expansion of human and livestock populations over the past four to five decades has placed extraordinary pressures on forest resources, undermining the nation's rich biodiversity. By the early 1990s, an estimated 27 per cent of all plant species and 65 per cent of bird species were classified as endangered.[2] In addition to the degradation of public forests, other common property resources (CPR), which were providing an important supply and range of biomass products to local communities, have dwindled. A study conducted in eighty-two villages in six Indian states, found that

CPR forest and pasture area had declined between 31 and 55 per cent since the early 1950s. Further, common lands have also lost species diversity and vegetation density.[3]

While the growth of human and livestock populations has long been identified as a major factor driving deforestation, as Westoby notes, 'It is an over-simplification to regard deforestation as the consequence of population growth. It is nearer the truth . . . to regard deforestation and population growth as joint manifestations of exploitative social relations.'[4] From the colonial to the post-Independence era, government policies that encouraged state control and commercial exploitation of forests and common lands have contributed substantially to deforestation and ecological deterioration, through both direct felling and the destabilization of indigenous management systems.[5]

With the destruction of the larger biological systems upon which India's agrarian economy ultimately depends, the smaller economic subsystems of both commerce and rural subsistence collapse correspondingly. While the government of India has invested billions of dollars in reforesting degraded lands with fast-growing, exotic tree species, complementary strategies, which can help stabilize and restore disturbed natural forests, must also be developed. Natural forest regeneration offers encouraging signs that a range of low-cost, replicable options can help to biologically restore India's degraded lands. To further explore the response and power of nature in resuscitating badly disturbed environments, and to objectively assess natural regeneration as a viable resource management option, a more comprehensive ecological understanding of both degeneration and natural regeneration processes is needed.

This chapter analyses the definition, extent, and condition of India's degraded lands. It reviews factors contributing to degradation and their ecological impact, and compares social forestry schemes with natural regeneration options for degraded lands, analysing their potential and merits. The chapter concludes with strategies to encourage natural regeneration management initiatives.

Defining Forest Degradation

According to Blaikie and Brookfield, 'Land degradation should by definition be a social problem. Purely environmental processes

such as leaching and erosion occur with or without human interference, but for these processes to be described as "degradation"
implies social criteria which relate land to its actual or possible
uses.'[6] Consequently, land or forests are considered degraded when
they suffer a loss of intrinsic qualities, which results in a decline
in their capability to satisfy a particular use. Deforestation may
not necessarily be considered degradation, even though changes
in micro-climate, hydrology, and soil take place, since these lands
may be brought under productive agriculture. It is only defined
as such 'if the degradation process under one system of production
has reduced the initial capability of the land in a successor system
. . . In actual practice, this is often the case, since more serious
degradation reduces capability for most, if not all, future possible
land uses.'[7] Certainly, in India, where much of the remaining
forest lands are located on soils with marginal fertility, incapable
of supporting productive agriculture, deforestation often undermines the productivity of the original forest ecosystem, as well as
the capability of the land to support alternative uses. The vast
areas of the once dense, old-growth forests that now exist as
unproductive, devegetated tracts devoid of topsoil, demonstrate
this point. Since human activities and natural processes can both
contribute to degradation and restoration, Blaikie and Brookfield
suggest that the equation for net degradation reflects the outcome
of natural degrading processes plus human interference minus
natural reproduction and restorative management.[8]

A common method for assessing forest degradation is to com-
pare the current state of a specific ecosystem with its former
undisturbed condition. Even within a single forest ecosystem,
species composition, structure, and density patterns will vary
according to highly localized moisture levels, slope, aspect, and
other variables. Consequently, the extent of degradation should
be measured against the original state of the forest in specific
contexts prior to the initiation of intensive disturbances by human
and domesticated animals. This presents serious methodological
problems for accurately assessing changes in forest ecology. While
it is impossible to assess the precise patterns of ecological degeneration that has occurred in a given area, it is possible to
observe the general patterns regarding biological changes that
have and continue to take place.

In India, ecosystems experiencing moderate to serious disturbances are classified as wastelands. Degraded forests are typically defined by the government as land with less than 40 per cent crown cover. While providing a useful indicator of canopy closure, this definition of degradation does not reflect other changes that may occur underneath the crown, affecting soil fertility, moisture availability, species composition, and general biotic productivity. There is a need to develop more sensitive measurements of forest degradation to better monitor ecological change. Nonetheless, in forest areas with diminished tree canopy, a dominance of thorny, bushes or devegetated, rocky subsoil are commonly found.

Extent of Deforestation and Degraded Land

According to the National Remote Sensing Agency, the annual average deforestation rate in India between 1975 and 1982, was 1.3 million hectares, representing the degradation of 10.4 million hectares of closed forest.[9] Although the Ministry of Environment and Forests reports that Landsat data indicate an increase of 56,000 hectares of forest cover during 1989–91 it is likely that this is due to the massive plantation programmes. Many field observers, however, note that much of India's natural forests are continuing to degrade.[10] Although the decline in deforestation rates in India is a welcome and significant reversal, the distinction between deforestation and degradation is an important one. Satellite imagery often fails to pick up the more gradual processes of ecological decline. In fact, it is this successive deterioration of vegetative diversity, forest density, structure, and regeneration potential that poses the greatest threat to India's natural forests. Unless the process of degradation can be arrested and reversed before natural resilience is lost, opportunities to rehabilitate India's degraded forest lands will be seriously jeopardized.

The Society for the Promotion of Wastelands Development estimates that in 1984, 35.9 million hectares of forest land and 94 million hectares of non-forest area — both private and communal — were degraded.[11] This is equivalent to 40 per cent of the subcontinent's total land area of 328 million hectares. The process of devegetation associated with the creation of wastelands also leads to accelerated soil erosion, resulting in the loss of 6000 million tons of topsoil that is washed or blown away annually.[12]

Processes of Forest Degeneration

Forest degradation has numerous agents and causes, many of which reinforce the process. In India, human activities contributing to the degradation of natural forest vegetation, frequently involve five primary processes: (1) removal of woody species at rates exceeding their renewability, (2) excessive grazing by livestock of grass and other herbaceous matter, (3) increased incidence of fire, (4) topsoil loss and soil compaction, and (5) a decrease in water retention and recharge capacity. Degradation often occurs gradually as a succession of biotic disturbances slowly deplete vegetation, suppressing natural regeneration, and inducing sheet and gully erosion of topsoil. This type of degradation process may occur over several decades. Figure 10.1 depicts the more gradual process of vegetative changes common in many regions of India.

Degradation of forests often begins with the selective logging of larger trees, causing openings in the crown cover. Timber extraction results in damage to smaller trees, saplings, shrubs and herbs, as well as causing soil erosion and compaction. Roads cut into forests to facilitate logging open the once-isolated forests to other outside users. Further felling of trees can cause larger losses of crown cover, opening up more space for livestock grazing. In turn, grazing frequently causes further soil erosion and compaction, suppressing natural regeneration through loss of soil fertilizers and both seed and seedling destruction. Progressive losses of plant diversity and grass productivity are often coupled with the invasion of exotic, non-palatable weed species. Where fuelwood is scarce, local people may uproot stumps and roots, further disrupting soils and accelerating erosion. Ultimately, overexploitation through cutting and grazing may cause a near total loss of vegetative cover, topsoil, and soil nutrients. The ecological impacts of different anthropogenic interventions are summarized in Box 10.1. With each biotic interference, the critical impacts are relatively similar: loss of species diversity; disruption of nutrient cycling and loss of topsoil and its nutrients; progressive suppression of natural regeneration from both coppice and seed; and decline of biomass productivity and balanced ecological functioning.

In the end, the complex multi-tiered, multi-species forest structure is transformed to a depleted, highly vulnerable biological system, with little or no vegetative cover save for some scattered, thorny and non-palatable species. The soil structure is also altered — from deep topsoil, high in micro-organisms, humus, and clay or silt, to a pauperized condition with total loss of humus followed by topsoil erosion. Declining moisture-holding capacity of topsoil, accelerating surface water run-off, and minimal groundwater recharge capacity are associated with degradation.

Reversing Forest Degradation: Options and Strategies

The opening of the forest canopy also results in a reduction of critical fungi living on or in close association with the roots of many plants.[13] Mycorrhizae facilitate the absorption of nutrients and moisture for many plant species. Consequently, these microscopic, below-ground organisms play an important role in a large variety of plant communities, facilitating the process of forest succession in degraded ecosystems. Research indicates that the removal of canopy trees results in high substrate temperatures, which constrain the growth of mycorrhizal fungi, in turn reducing seed survival.[14] Deforestation also disrupts the food supply to many micro-organisms, causing their eventual death. 'When this happens, soils become virtually sterile and inhospitable to both higher and lower forms of organisms . . . The re-establishment of the mutualistic relationship of mycorrhizal fungi and host plants is the key to any restoration activity . . . particularly where trees are being re-introduced to degraded sites.'[15]

India's rich forests have gradually been reduced over the centuries. According to Norman Meyers, as much as 90 per cent of the subcontinent's forest cover has disappeared.[16] The total area under forest cover in relation to India's total land mass has declined from nearly 17 per cent in 1972, to 14 per cent in 1982, to approximately 11 per cent in 1992.[17] The shrinking forest cover stands in sharp contrast to an ideal forested area of 33 per cent.[18] As problems of forest degradation have escalated and become increasingly recognized over the past two decades, the government of India has initiated a number of programmes to rehabilitate disturbed ecosystems. These efforts have been carried out in conjunction with stricter controls on commercial timber harvesting, the extension of fuel-efficient stoves and biogas programmes, and attempts to substitute both non-forest raw material and finished products.

Plantations and Conventional Social Forestry Programmes

During the mid-1970s, concern over the rapid forest loss and the declining availability of wood products led government planners to introduce the concept of social forestry under its fifth Five-Year Plan (1976–81).[19] The original goals of the social forestry programme were to produce firewood, fodder, and small timber resources to meet the biomass needs of local communities, while helping relieve pressures on natural forests. The scheme aimed to reclaim wastelands through afforestation programmes that would generate employment opportunities in rural areas and restore the ecological balance. The social forestry programmes were infused with added attention and momentum through a large influx of funds from foreign governments and international agencies. Centrally sponsored social forestry projects in virtually all states were initiated during the early 1980s. Even commercial banks and industries developed programmes to provide assistance to farmers to establish tree plantations for specific purposes. The social forestry projects included community woodlots, strip plantations and farm forestry, and focused primarily on communal and private lands. The concept implicitly assumed that most natural forests would remain under strict forest department control to meet industrial and commercial needs, while common and private lands would be reforested through social forestry schemes to respond to the fuel and timber needs of rural communities.

Ecological Impact of Social Forestry Plantations

Between 1980 and 1990, social forestry activities expanded rapidly. The programmes achieved dramatic success particularly on private farmlands in the Punjab, Haryana, western Uttar Pradesh, and Gujarat — India's 'green revolution' belt.[20] However, rather than planting for subsistence fuelwood and timber requirements, farmers and communities began selecting fast-growing species, especially eucalyptus (hybrids of *E. tereticornis*), to respond to high market prices and labour shortages. In Uttar Pradesh, farmers exceeded the original target of 8 million seedlings, planting 350 million between 1979 and 1984.[21] In the Punjab, in only a decade, more than 3 per cent of the cultivated area had been planted under eucalyptus.[22] In Gujarat, from 1983 to 1984, farmers planted 195 million·trees, quadrupling the 49 million mature trees previously targeted for the state.[23] Studies estimate survival rates for trees between four and eight years of age between 60 per cent[24] and 77 per cent.[25] While social forestry programmes have resulted in the establishment of billions of trees — mostly exotic, fast-growing species — most farmers and participating communities planted them with the expectation that they would be harvested in six to ten years and sold in commercial markets. The enthusiasm of rural people 'to plant eucalyptus declined after 1986, as the tree failed to generate the kind of returns that farmers were expecting from its sale. Some farmers removed the tree roots and went back to annual crops.'[26] Consequently, while social forestry–subsidized eucalyptus plantations have allowed India to reduce the annual rate of deforestation estimated at 1.3–1.5 million hectares per year between 1980 to 1985, to approximately 50,000 hectares per year, it is questionable whether these newly established woodlots will be sustained.[27] Since most social forestry projects are designed around short rotation felling, unless carefully controlled, this may reinitiate the degradation process. Many cut-over plantations may re-emerge as large areas without tree cover, which will be further subjected to degradation through a host of consecutive biotic disturbances.

The conventional plantation approach has led to serious losses of local tree diversity on common and private farmlands. Monoculture plantations with tightly spaced, high-density trees suppress natural regeneration and vegetative undergrowth due to shading and competition for soil nutrients and water. In contrast, the undergrowth and multiple-storey vegetation in mixed cultures and natural forests encourage a range of micro-habitats. In plantations, due to the loss of diverse, multi-tiered floral micro-niches, faunal diversity also declines. Loss of plant diversity additionally deprives local communities of a whole range and continuous flow of plant products such as food, fodder, oilseeds, leaf manure, and raw materials for artisans. Exotic, monoculture plantations have also proven substantially more vulnerable to pest attacks and disease. The lack of a multi-tiered structure in plantations also reduces ground cover and leaf litter, critical for reducing soil erosion and slowing water run-off. Raising short-rotation plantations of monoculture species like eucalyptus, which is repeatedly harvested, leads to depletion of soil nutrients. Subsequent harvest yields progressively decline. Important ecological implications of the conventional block plantation approach are illustrated in Box 10.2.

In short, while India's experiences with farm and community tree plantations have been relatively successful in responding to market demands for construction-poles, pulpwood, and timber, they should not be equated with the restoration of natural forests which serve very different economic and ecological functions. Ultimately, short-rotation plantations of monoculture species should be viewed more as a form of agriculture than as a management option for forest ecosystems. Although more than a billion dollars has been invested in social forestry plantation schemes in recent decades, they have failed to slow the process of natural forest degradation. Further, a huge area of wastelands needs restoration in India, encompassing up to one-half of the subcontinent's terrestrial landmass. Thus, it is not realistic to expect that enough human and financial resources will be available to restore this land. With plantation costs averaging Rs 10,000 per hectare, billions of dollars would be required to artificially replant a sizable proportion of India's degraded lands. Furthermore, the very low productivity of most forest plantations does not justify the current high costs incurred on restoring degraded natural forests, nor are markets available to absorb the produce. The biomass needs of village communities can be met better through the production of a large diversity of plant species, not by tree species alone. While plantation forestry will continue to be an important strategy to generate raw materials for plywood, pulp industries, packing cases, and to provide fuelwood and some construction timber, alternative management strategies need to be developed to restore India's degraded natural forests.

The strong emphasis of conventional social forestry afforestation programmes on degraded common and private lands has diverted attention and support away from the management problems of state forest lands. Further delays in reversing the process of forest degeneration will finally render the land ecologically unfit for rapid revegetation and rehabilitation. The operational problems faced by forest officers attempting to implement social forestry projects in the face of pervasive degradation over large territories are exemplified in Box 10.3.

There is an urgent need for a strategy that will permit rapid reclamation. The strategy must offer a diversity of forestry options that are compatible with physical factors such as soil and rainfall, as well as desired product outputs. Natural regeneration under community protection provides a low-cost alternative to plantation restoration.

Natural Regeneration Patterns

In recent decades, natural regeneration has received inadequate attention from foresters. The promotion of social forestry plantations has preoccupied forest managers, planners, and field workers, drawing attention away from the opportunities that natural regeneration presents. Natural regeneration provides an attractive and widely replicable alternative to reclaim degraded lands quickly to enhance their biological potential, while meeting the diverse biomass needs of local communities in a sustainable way.

Natural regeneration comprises processes of secondary forest succession. The level of plant biomass depends upon the availability of sunlight, water, and mineral nutrients, as well as the impact of factors such as floods, grazing, and other human interventions. The natural process of vegetative succession involves the gradual replacement of ecological phases — low biomass vegetation dominated by grasses and herbs, to shrubs and young, pioneering woody saplings, to a state of high biomass dominated by tree canopy. Rates and quality of regeneration are determined both by the natural characteristics of the local ecosystem (e.g. rainfall, altitude, slope and aspect, soil type) and its level of degradation.

If·protected before sites become seriously disturbed, both natural forest regeneration and 'enhanced' or 'assisted' natural regeneration have proved to be effective methods to reverse the degradation affecting natural forests. In many cases, natural regeneration will occur rapidly if abused lands are provided simple protection from anthropogenic disturbance. Social fencing based on co-operative community agreements and actions are often effective in protection. Access controls are normally imposed by communities who establish patrols to stop grazing and cutting. Once regenerative processes are under way, community management groups may decide to allow some thinning or grazing. Initially, logging, fuelwood cutting, burning, and grazing must be stopped or lightly controlled. Even if it proves impossible to halt grazing, experience from the Western Ghats and West Bengal indicates that grazing must at least be prevented during the first two to three years of recovery. Some communities bring degraded forest land under protection gradually, facilitating the initiation of regeneration in certain areas first, while other areas remain open for grazing and collection. Once regeneration advances and tree saplings grow above grazing levels — averaging 1 metre to 1.5 metres — these lands can be reopened as regulated pasturage. Sequentially, over a period of two to four years, villages can bring increasing amounts of land suitable for natural regeneration under controlled access.

Forest Succession and Natural Regeneration

The process of natural regeneration also follows a succession pattern, characterized by a series of phases that vary in length and type depending on the environment and degree of disturbance experienced previously. Figure 10.2 depicts the process of natural regeneration under two different scenarios: a degraded forest environment with high coppicing potential, and a forest with few coppicing species, but good tree-seed supplies. In areas with high coppice potential, some silvicultural operations may be required to clean the rootstock. During the first rainy season, coppice shoots, grass, shrub, and herb vegetation will emerge. In the next few years, shrubs and saplings dominate, leading to the decline of the grass-herb layer. The coppice shoots continue to grow; thinning of coppice shoots will accelerate stem growth. In five to ten years, the coppice stems will develop an increasingly closed canopy, suppressing even the shrubs. The trees will reach a height of about 10 metres after fifteen years, and some species may be ready for harvest as construction poles. Others will start producing non-timber forest products (NTFPs), and in twenty years, tree growth rates may decline. After long periods of protection, climax species may appear and the tree growth rate would be nearly zero.

In many situations, such as in the Western Ghats or Haryana, coppice potential is low because of the composition of forest species. In areas with a low percentage of coppicing tree species, but an ample supply of mature, seed-dispersing trees, natural regeneration is likely to require a modified time frame and follow a different pattern than trees in a coppicing area. With protection, grass production will increase with herbs starting within the first year. By the second or third year, the soil moisture may improve and germination of tree seeds from natural seed dispersal mechanisms (wind, birds, bats) is likely to occur. By the fifth year, grass production may peak, and shrubs may start to grow; tree seedling growth will continue at a slower pace. The tree seedlings may reach a height of 5 metres by ten to fifteen years. By that time, the canopy would be closing, grass production would be marginal, and shrubs would also decline. By the twentieth year, long gestation trees will start yielding NTFPs (e.g., mango, *jamun*, and tamarind) and some species may start even earlier. Certain coppice species would be ready for harvest. Very long periods of protection may lead to changes in species composition from deciduous to evergreen as observed in the Western Ghats.

In extreme situations, where soil erosion has reached conditions characterized by gullies and ravines with little or no vegetation, natural regeneration alone may have little immediate impact in improving vegetative cover. In such cases, alternative, highly intensive forestry options are required. The following review highlights experiences with natural regeneration in the semi-arid Shiwalik hill tracts of Haryana, in the sal coppicing forests of south-west Bengal, and in the moist evergreen forests of the Western Ghats of Karnataka.

Experiences with Natural Regeneration: Case Studies

The following three case studies document changes in biodiversity and productivity after varying periods of community protection and management in three distinctly different ecological zones. Open access control plots receiving no form of management were selected to assess the impact of uncontrolled use on vegetation.

The parameters monitored include species richness, species distribution, tree density, size of trees, basal area, grass productivity, and extent of coppicing. Vegetation was monitored in sample quadrats and data were recorded for tree, shrub, herb, and grass layers.

In all study areas, promotion of natural regeneration was possible only through the participation of the local community; little or no capital or technological investments were made. In the Karnataka study-sites, protection and management was undertaken solely by the community; in the two other states, management was jointly implemented by the community and the forest department. Both institutional systems provided viable opportunities to regenerate degraded forest lands.

In all locations, prior to community protection, vegetation was very poor, with no trees and only remnant shrubs and stumps. The soils were eroded, and the community had free and unlimited access to harvest or graze. There was no management system in operation, which could control extraction of biomass, overgrazing, and progressive degradation.

The Evergreen Forests of the Western Ghats

Sagara block in Karnataka's Western Ghats was selected to represent natural regeneration under community protection for both deciduous and evergreen forest ecosystems. The area receives an annual rainfall of about 100 centimetres. In the study areas local initiative led to the promotion of forest protection activities to facilitate natural regeneration. Initially, grazing was banned for three years, but later free grazing was resumed. No harvest of green wood was permitted. Only fallen twigs could be harvested with the village committee's permission. Although there was no official guard, local access control was strictly enforced by the entire village. In Hunsur village, the forest was guarded by each household for one day on a rotational basis. Any harvest or removal not permitted by the village committee was punished by imposing fines. In extreme cases of forest abuse, a social boycott was imposed. The Karnataka Forest Department was not involved in this self-devised community management system.

In two villages, Padauagodu and Honkeri, the plots under protection have regenerated for ten years; and in Alahalli, plots have been under protection for fourteen years. Hunsur, a village with a long tradition of over 100 years of community management, was also selected. To ensure similar socio-ecological conditions, all four study villages were within a radius of 10 kilometres. In each study area an unprotected control plot adjacent to the village was chosen, allowing a comparison of natural regeneration in protected and unprotected patches.

The study indicates that the number of different tree species over 1.5 metres in height increased from zero in the unprotected control plots to twenty-six after fourteen years of protection in Alahalli. The species diversity rose to fifty-five in the 100-year-old protected forest of Hunsur. The number of trees larger than 10 centimetres in diameter at 5 feet above the ground (GBH) also increased from zero in the unprotected plot to 1477 after fourteen years of protection. The number of shrubs and herb species also increased up to fourteen years of protection; these numbers appeared to decline thereafter, possibly due to the growth of tree canopy and shading.

Basal area provides an important indicator of standing tree biomass or standing tree stock. In the control plots, there was no standing tree biomass. In contrast, in plots under protection, standing basal biomass per hectare increased from 4.1 square metres within ten years of protection to 9.7 square metres within fourteen years of protection and finally to 33 square metres in the century-old Hunsur site.

Succession from deciduous to evergreen species could also be distinguished when comparing the dominant species in the fourteen-year-old and 100-plus-year-old plots. The findings clearly show that natural regeneration under community management has led to significant increases in species diversity, tree density, and biomass. In Alahalli village, only 12 per cent of the tree regeneration was based on coppice growth, while in Honkeri village, more than 50 per cent of the trees originated from coppice rootstock. This indicates that given adequate seed sources, even in the absence of coppice rootstock, natural regeneration by seed can succeed.

The Sal Forest of East Midnapore

East Midnapore forest division of West Bengal was selected for monitoring vegetative changes. The area receives a mean annual rainfall of 120–150 centimetres. Sal dominates the vegetation on generally poor laterite soils. The promotion of natural regeneration was initiated by the forest department under a joint forest management system. The programme involved the formation of Forest Protection Committees (FPCs) in each village. FPCs entered into an agreement with the forest department to jointly protect and promote natural regeneration in a patch of degraded sal forest. Each participating community was assured 25 per cent of returns from sal tree harvests and was guaranteed full access to all NTFPs. FPCs generally did not permit grazing initially; however, some grazing was permitted as the forest recovered.

In the Moupal-Ranja beat of east Midnapore, three forest plots near a village were selected for study. The plots included one unprotected control plot of degraded forest, a sal patch protected for five years, and, in a neighbouring village, a sal patch protected for ten years. The number of trees above 10 centimetres in GBH increased from zero in the unprotected site, to 765 after five years and 961 after ten years. Correspondingly, the basal area increased from zero to 7.4 square metres per hectare after five years, and to 16.5 square metres per hectare after ten years, with an annual growth rate of 1.6 square metres per hectare. The number of shrubs, herbs, and climbers increased sharply whereas the lower-storey species declined somewhat as the canopy closed, presumably due to shading by the sal trees. The number of climbers, whose tubers are a crucial source of food for the locals, nearly doubled during the first five years of protection to 235 per hectare.

The Mixed Deciduous Forest of the Shiwalik Mountains in Haryana

The study locations were selected from the Morni and Pinjore forest ranges of Ambala division. The area is hilly with steep ravines and a mean annual rainfall of 100–120 centimetres. Natural regeneration was facilitated by protection provided by community-based Hill Resource Management Societies, which were formed to promote soil and water conservation, and manage the eroded Shiwalik watersheds. Later, a joint forest management system was introduced. Grazing was banned in the selected area and protection was done voluntarily.

In Haryana, researchers investigated the processes of natural regeneration and its effects on grass production under varying periods of protection. Sample plots were chosen to reflect three, six, and ten years of protection from grazing. A set of unprotected control plots subjected to free grazing was selected for comparison. The researchers found that the number of trees per hectare increased from 90 in the unprotected plots to nearly 500 after ten years of protection. The number of shrubs also increased during the first three years, and then declined slightly, possibly due to an increase in trees and shading. Grass productivity increased from less than 1 ton per hectare per year to nearly 3 tons during the initial three years of protection. However, per hectare grass yields, declined by nearly 50 per cent between three and six years of protection, falling somewhat further up to ten years of protection, probably due to the increasing number of trees and shrubs shading out lower-storey grasses. While natural regeneration has clearly led to increasing biomass and basal area in protected areas, tree and shrub vegetation in regenerating plots may need to be manipulated to ensure a steady supply of fibre and fodder grasses to communities.

The initial findings from three diverse ecological contexts in south-western, eastern, and northern India demonstrate that community protection has led to significant increases in plant diversity and biomass production. Yet, evidence supporting the efficacy of natural regeneration for restoring degraded forest lands is much stronger. In West Bengal, over the past decade, 350,000 hectares of degraded sal forests have been brought under the protection of 2350 community FPCs.[28] The impact of local management on forest regeneration is clearly visible by analysing a time-series of satellite images of the region, as well as from field-level observation. Thousands of village forest protection groups operating in neighbouring southern Bihar and Orissa have had similar success in regenerating degraded forests with little or no capital investments. In the Sarangi Range of Dhenkanal Division in Orissa for example, more than 100 FPCs protect several thousand hectares of once badly disturbed forests. In Rupabalia forest, the community noted that prior to protection the hillside forest above their village had been reduced to a stony wasteland with little grass. After fifteen years of protection the hill is covered by a dense, multi-storyed forest with high diversity of trees, shrubs, and herbs.

More studies are required to identify patterns of vegetative changes in other parts of the country under natural regeneration.

High Priority Areas for Natural Regeneration

In her pioneering study of degraded forests in India with high natural regeneration potential, Manjul Bajaj used information on species composition, soil type, rainfall, and climatic conditions, combined with land satellite imagery and ground surveys indicating degraded areas.[29] The research found that forests with species that are resilient coppicers (sal, *dalbergia, khair,* dhak, and *anogeissus*), fire resistant (sal, pine, *sisso*), grazing resistant (*Butea monosperma-dhak,* acacias, *Zizyphus*), and colonizers (*adina,* pine, *alnus,* and *dalbergia*) had high regenerative potential. Areas with rainfall above 100 centimetres per year and annual minimum temperatures greater than 10 degrees celsius, were also considered conducive environments for natural regeneration. Bajaj found that these conditions prevailed in about 63 per cent of India's forest lands, covering over 40 million hectares. Further, ground studies and remote sensing data indicated that of the 40 million hectares, 29.5 million hectares were degraded with less than 40 per cent canopy closure and in need of community protection if their high natural regenerative potential were to be unleashed.[30] Another recent report noted that degraded forests with less than 40 per cent crown density, occupy 32 million hectares, constituting nearly 46 per cent of the nation's forest area. Much of this area has good ecological resilience and could be targeted for diagnostic assessments and support programmes, especially where communities are already protecting or motivated to assume protection responsibilities. Much of the remaining forests with 40 per cent or greater crown density may have already suffered extensive disturbance and may be rapidly degrading. Consequently, these areas will also need better access controls and require community protection.

More precise field assessments would allow a finer definition of priority areas for natural regeneration. The sequential stages of degradation shown in Figure 10.1 need to be assessed in determining the viability and rapidity of natural regeneration in different forest environments. In stage 3, where logging has occurred but some remaining trees and good rootstock exist, natural regeneration potential is high. In stage 4, although no trees may be present but good coppicing stumps exist, natural regeneration also offers opportunities for rapid regrowth, especially if accelerated by modest soil and water conservation measures. By stage 5, coppicing tree stools may have been removed, topsoil losses are severe, and the hydrological balance is disrupted; weedy species dominate, undermining the establishment of more valuable local species. In stage 6, where most vegetation and topsoil are lost and hydrological conditions are poor, natural regeneration may be impossible, or at most a very slow process, requiring intensive capital investments in soil and water conservation, and plantation. There is an urgent need in India to begin field-level mapping of high potential areas, particularly those in stages 3 and 4, where communities are interested in managing forest resources. Finer spatial definition of high potential regions for accelerating joint forest management initiatives will help guide the forest department and foreign donor investments in this sector.

Approaches to Sustainable Forest Management

Although stopping degradation and initiating natural regeneration through community protection is the first step in restoring the productivity of forest ecosystems, natural systems can be manipulated positively to accelerate recovery and enhance the flow of desirable products. Human interventions, however, need to be directed by specific management objectives.

Soil and water conservation measures such as contour trenching, vegetative bunding, and small check-dams can enhance soil moisture and the accumulation of topsoil, accelerating rehabilitation of the micro-environment. Such conservation measures are helpful in improving germination and growth rates of seedlings. Enrichment planting of desirable local and exotic tree species in degraded forest gaps can generate additional forest products and improve forest density and composition. Cleaning degraded forest lands of dead brush, cleaning stumps, and cutting excessive coppice shoot growth can also facilitate regeneration and promote healthy growth. More extensive enrichment planting is suitable for locations with no coppice or low coppice potential and with moderate levels of soil degradation.

Natural forests can also be manipulated to generate different products, depending on the needs of communities. Artisan communities may place greater importance on specific raw materials needed for their cottage industries. Rope-makers require fibre grasses; potters need large quantities of fuelwood for their kilns; women involved in disposable plate making use sal and dhak leaves in large quantities. Forest resource requirements of different agricultural and pastoral systems also vary. Some farming systems, particularly in the Himalayas and the Western Ghats, depend on heavy nutrient transfers in the form of leaf mulch collected from forest lands. Pastoralists need large supplies of fodder like grass or leaves; forests may be used for open grazing or cut-and-carry systems, depending upon the size and composition of cattle. In other farming systems, the role of the forest may be most important in stabilizing micro-climatic conditions and hydrological functions, ensuring a steady supply of wood for farming tools, and maintaining a habitat for birds that control insect pests. Rural women often require the availability of a diverse range of forest products for food, medicines, tools, fuel, and fodder. One study indicates that certain multi-purpose trees, like *Zizyphus* and *Prosopis,* are particularly important to women. These species tend to be low, versatile coppicers with large crowns, which generate fruits, fodder, and twigs and leaves for fuel.[31]

If manipulated and managed carefully, natural forests can substantially improve the productivity of desired species, absorbing labour while generating greater income in cash and in kind according to seasonal demands. Unfortunately, only limited scientific research has been conducted to identify the most effective ways to manage natural forests intensively for both timber and non-timber products. While community-based, ethnobotanical knowledge is extensive for some ecosystems, it is often undocumented, and rarely used by outside planners. Further, communities have neither the tenurial security nor the right to manipulate forests for long-term productivity increases. In the future, if the natural forests of India are to be intensively managed to optimize yields of desirable products, a broad-based programme of systematic experimentation will need to be initiated jointly by communities and forest departments, in collaboration with universities and non-governmental organizations.

Strategies to Optimize Sustainable Productivity

Increasingly, foresters are beginning to think in terms of managing ecosystems rather than managing a few valuable timber species. Ecosystem management is more complex, because manipulations in one part of the system often affect other components. If forest management is to adapt itself to respond successfully to community objectives, new systems of production will need to be designed, tested, and monitored. The development of non-timber forest products must emerge as a major forest management strategy. NTFPs include fodder, fruits and other food, fibre, gums and resins, oils and aromatic extracts, tannins, medicinal plants, animals and insects, structural woody material, religious and ornamental articles. In other words, NTFPs exclude poles, logs, and pulpwood. From plants, NTFPs may be derived from the leaf, flower, fruit, seed, twig, pods, stem, roots, tubers, or bark. Unfortunately, not much is known about the practical ecological requirements of most NTFPs, including their management within a multiple-use forest, yield-data for different NTFPs, varietal and genetic improvements, and traditional patterns of manipulation and collection.

Reorienting management emphasis to multiple-use, non-timber forest production will help maximize both social and ecological benefits. Unlike timber-harvesting which typically causes major ecosystem disturbances, NTFP collection tends to be much less destructive. If product extraction rates do not exceed sustainable yields and technologies remain simple, the harvesting of timber and non-timber products that are annually renewable may be relatively benign and ecologically sustainable. Furthermore, having used NTFPs for centuries, tribal communities have significant knowledge about their uses, propagation, and ecological requirements.

NTFP-based systems have other advantages. Since they often involve optimization of a diversity of products, they contribute significantly to the conservation of biological diversity. Leafing, fruiting, and flowering patterns are typically seasonal, often becoming available during periods of food shortage or high unemployment, providing important labour opportunities and supplementary income and products during non-agricultural periods. NTFP collection activities tend to be labour intensive, small, and often household-based. Perhaps most significant, because their opportunity costs for collection are the lowest, NTFP systems are disproportionately accessible to the poorest, most forest-dependent minority groups, including tribals, landless families, and women. In fact, harvesting and management of NTFPs are most often practised by rural women; consequently, product benefits, whether they be cash or subsistence items, are likely to be more equitably distributed, both within communities and households. These characteristics combine to make NTFP systems highly strategic for the local economy and the local environment.

Management objectives must specify desired ecological and economic outcomes, reflecting the structure and composition of the forest, as well as indicating the level of product flows desired and the periods when they are most needed. In the forests of south-west Bengal, for decades the forest department has felled the sal trees every six to fifteen years to maximize revenues from the sale of construction poles. The primary goal of this management strategy was to maximize pole yields; however, under this short rotation the sal trees never mature to begin producing valuable oilseeds or reach a size where timber can be harvested. Further, by relying on optimizing sal pole production, markets have been flooded, depressing prices. Average prices received by the Arabari farmers in west Midnapore for 3-inch-diameter poles fell from Rs 14 in 1985 to Rs 5 in 1989.[32] Considering the oversupply of poles, more diverse product management systems need to be developed to maximize and stabilize income flows.

The clear-felling of the sal on a short rotation disrupts the forest ecology and many useful associated species. It also disturbs longer term succession patterns that renew soil fertility. A.N. Chaturvedi contends that rapid rotation fails to provide sufficient opportunities for humus accumulation, soil mycorrhiza development, and regeneration through seeding. He suggests that many sal coppice forests should be used primarily for NTFP collection and fuelwood thinning, allowing them to mature into primary forests over a period of eighty to hundred years.[33] When the forest is felled, villagers report that many important products disappear, while micro-climatic changes occur including the loss of localized humidity and the presence of stronger, hotter winds. To meet both the need for poles and to better maintain the natural ecosystem, the forest department and communities are discussing rotational harvesting of very small patches of forest, 1 hectare to 5 hectares, to protect the healthiest sal to mature into 'mother' trees. Alternatively, some communities may want to gradually thin out coppice growth to allow sal and other useful trees to mature, thereby facilitating the growth of valuable grasses, herbs, and shrubs.

Thinning, cleaning, soil and water conservation, enrichment planting, and timing harvests can all be used to facilitate growth of certain species, and increase and stagger productivity flows. Thinning and pruning allow more light to fall on desired species, accelerating their growth. Cleaning removes dead and decaying organic material, providing room for new shoots to emerge and stimulate healthy coppice stem development. Unfortunately, much of India's forests are not managed in this manner. Forest departments rarely have the budgetary or staff resources to carry out these tasks on more than a small fraction of their forest lands. Communities have no legal authority or usufruct security to manipulate forests. As a result, the productivity of most of India's natural forests is far below their potential.

For example, in south-eastern Gujarat, the state forest department planted thousands of bamboo culms over the past ten years.[34] The bamboo (*Dendroclamus strictus*) was well-suited to the environment and quickly established itself in the degraded natural teak forests near Limbi village. Yet, the productivity and quality of the bamboo have been far below its potential due to the dense buildup of dead leaves and other organic material. The abundance of litter within the culm has suppressed the growth of new shoots and poses additional fire hazards during the dry season. If the stands were routinely cleaned and thinned, the danger of fires would be reduced, productivity would increase, and a regular flow of bamboo poles would be ensured. Since many *Bhanjara* basket makers live in the community, the demand for bamboo is high. Unfortunately, no mechanism exists to allow villagers to manage the bamboo culms, and the forest department cannot conduct cleaning operations without a budget to hire labourers. The management problem could be resolved by formulating an agreement to allow villagers to clean the bamboo routinely, giving them rights to the thinnings in lieu of pay. Villagers could even be encouraged to fertilize the culms to increase the bamboo's productivity and the community's profits.

Another example of new approaches to management is being developed by foresters and villagers in Harda Forest Division in Madhya Pradesh. In this 142,000-hectare forest tract, 155 village forest protection committees have been formed and now co-manage 70 per cent of the forest area.[35] Initial activities focused on grazing and fire control. Thousands of hectares of forests in Harda Division recently experienced profuse flowering of bamboos, an event that occurs every thirty-five to fifty years. Without protection from annual fires and heavy grazing pressures, bamboo regeneration after flowering is almost impossible. To allow the bamboo to become established, communities developed a system of rotational grazing, effectively keeping cattle out of flowering areas. Fire protection strategies have also been initiated by local forest protection committees, based on participatory rapid assessments where villagers analysed their own use patterns. Villager innovations include backfiring along all footpaths, since discarded bidis (Indian cheroots) at the edge of paths are a common cause of fires. People have also stopped lighting fires under mahua trees, a practice common in Central India to facilitate flower collection. As a result of people's involvement in protection, lush bamboo clumps are now becoming re-established in many protected forest patches.

Foresters and community groups in Harda are also beginning to develop more collaborative, improved management practices. Villagers have complained that multiple shoot-cutting to eliminate excessive coppice growth, was being done improperly and damaging the trees. They also noted that different tree species need to be pruned differently. More specialized techniques are now being developed, not only for teak but for other species including fruit trees. Local villagers with specialized knowledge are now familiarizing forest guards and other community members with these silvicultural techniques. The district forest officer has also noted that sharing agreements must be clarified prior to any operations and has given community members exclusive head-loading rights to all coppice shoot-cuttings.

Finally, in the Harda area many forest protection groups have complained that since the lucrative trade in tendu leaves was taken over by state-run co-operative societies, both yields and total income have fallen. Tendu (*Diospyros melanoxylon*) leaves are used to roll bidis, a low-cost cheroot widely smoked in India.

Tendu leaves are the most important NTFP in Madhya Pradesh and are a major source of employment in forest areas. Since the co-operative's budget for pruning labour is limited, most tendu plants are not pruned; consequently, the plants do not produce a healthy flush of young leaves, resulting in low productivity and a poor quality harvest. Further, because this task is often performed by outside labourers, tendu plants are often mishandled and damaged. The co-operative societies have also shortened the collection season, leaving many plants unharvested. The district forest officer has proposed that forest protection committees be given sole collection rights, and communities have responded very favourably. He has also recommended that they take responsibility for pruning and culturing the tendu plants in their areas. This allows the community to shift from being paid labourers to becoming share-holding managers, with the authority to develop their own culturing systems, including pruning and harvesting procedures.

Whether planning fire protection systems, sharing benefits from intermediate silvicultural operations, or controlling tendu leaf production and collection, the villagers of Harda are beginning to play an active role in forest management. They are moving from joint forest protection to actual joint forest management. Local foresters assist communities to clarify their objectives and discuss different silvicultural and management strategies by holding participatory planning exercises. If new options emerge for Indian forestry, they will be in villages like those of Harda, where the forest department and local communities are exploring new approaches to co-operative forest management.

In badly degraded natural forest environments, community members and foresters will need to plan for short- and long-term product flow requirements, considering ways to facilitate the recovery of lost soil fertility and enhanced micro-climatic and soil moisture conditions. Enrichment planting with nitrogen-fixing species and vegetative soil conservation measures may be initially advisable, followed by selective planting of more valuable trees and shrubs. These measures, however, are uneconomical, often costing Rs 8000 to 10,000 or more. India is rapidly gaining experience with much more cost-effective methods. In southern Rajasthan, foresters and villagers are experimenting with directly seeding local species, including acacia spp., dhak (*Butea monosperma*), *pongamea* (*Derris indica*), and mahua (*Maduca indica*), to complement the regeneration of native acacia, *Zizyphus, Lannia*, and *Boswellia*. Villagers collect the seeds in neighbouring forests and plant them in locations with better soils and moisture levels including along fence walls, in places where large stones were removed, in trenches and pits, within the vegetative (*tur*) fences, and behind small check-dams. Survival and growth rates have been good, often surpassing planted seedlings within one to two years. Further, costs range from Rs 300 to 600 per hectare.[36] Low-cost restoration strategies can liberate communities and foresters from depending on centralized bureaucracies and large projects, allowing them to initiate regeneration activities at any time.

Product flows change over time as certain species compete with others for light and nutrients. In Haryana, rope makers are concerned over declining fibre grass yields as *khair* (*Acacia nilotica*) tree canopies begin to close under community protection and grazing controls. Foresters and villagers need greater opportunities to experiment with enrichment grass planting and tree-thinning strategies to ensure that grass yields can be maintained, while allowing forest succession to continue. In the undulating Shivalik hills of northern Haryana, there is a diverse range of microclimatic niches determined by soil conditions, elevation, aspect, and topographic position, whether in a valley bottom, low, mid, or upper slope, or ridgetop. Each location has the capacity to optimally support a different mix of species. Some niches are better suited for grass growth, while others are better suited for acacias and hardy trees; some nutrient- and moisture-rich niches are capable of supporting a good growth of mangoes and other valuable fruit trees. To take full advantage of the capacities of each micro-environment, more detailed, site-specific management plans need to be formulated by local communities.

Optimal harvesting methods should take into consideration both the timing of product demands and sustainable extraction levels. Villagers may require products during periods of drought or unemployment during the agricultural off-season. Producers may also want to time the harvest to correspond to better market prices; yet, to ensure optimal productivity, harvesting may need to be timed to natural maturation cycles. Forest managers also need to assess how much of a given product can be taken without damaging the individual plant. For example, village makers of leaf plates have been criticized for undermining the growth of trees. Subsequent studies indicate that most women leaf collectors rarely harvest more than 5 per cent of a tree's leaf mass and that harvesting at this level will have little impact on growth rates.[37] Medicinal herbs, however, are often overexploited to extinction. Valuable species, particularly those with roots or stems containing a high level of a desired chemical, are systematically sought after, leaving behind a non-viable population. This is a particular problem in the Himalayan forests. In such cases, enrichment planting or culturing techniques need to be developed by user groups, where necessary, with extension assistance from outside agencies.

Ultimately, much can be done to intensify and diversify management systems for India's natural forests. Tighter access controls are a prerequisite. Beyond that, both modern science and indigenous knowledge can be used to enhance productivity substantially, depending upon management objectives. If communities are to take the lead in rehabilitating and intensively managing natural forests, they will require the authority to clean, prune, thin, enrichment plant, and harvest. This can be done with little outside capital assistance. Far more critical is the formal legitimacy needed for them to take action.

Conclusion

With continuing deforestation affirming the failure of conventional forest management systems, only a new paradigm for managing India's natural forests will ensure their healthy and productive existence into the twenty-first century. No longer can management objectives for forest lands rest on commercial extraction of a limited range of timber and pulpwood products to supply industrial demands. Neither can we realistically expect that establishing fast-growing tree plantations will halt or replace degrading natural forests. New priorities and objectives must supersede commercial interests, despite their historically powerful lobby. By empowering communities to establish access controls, the nation's threatened natural forests can be rehabilitated. Many communities will assume management responsiblities if forest productivity is targeted towards meeting local community needs, both economic and environmental. Given the diversity of local needs among forest-dependent users, specific 'micromanagement' by multiple objectives will be required. This will require community-based diagnostic studies and micro-planning that assess local needs, user groups, the productive capacity of the resource base, mapping of forest boundaries, and management interventions to enhance the desired range of biomass product flows.

Fortunately, there is now increasing documentation from numerous Indian states of alternative, creative community management approaches that are already well-established and witnessing early successes. Through development of a co-operative partnership between forest departments and community user groups, joint forest management strategies can facilitate natural regeneration and improve management of natural forest lands. As biomass-dependent communities organize themselves around the protection and management of degraded forest lands, impressive gains in revegetation will be visible in many locations.

Notes

1. Ann Clark and Prabhir Guhatakurta, 'The World Bank in the Forestry Sector: Policies and Issues in India', *Wasteland News*, 9(1): 25, August–October 1993.
2. Ibid.
3. N.S. Jodha, 'Common Property Resources: A Missing Dimension of Development Strategies', World Bank Discussion Papers no. 169 (Washington, D.C.: The World Bank), pp. 22–4.
4. Jack Westoby, *Introduction to World Forestry* (Oxford: Basil Blackwell, 1989), p. 45.
5. Madhav Gadgil and Ramachandra Guha, *This Fissured Land: An Ecological History of India* (New Delhi: Oxford University Press, 1992), pp. 239–45.
6. Piers Blaikie and Harold Brookfield, *Land Degradation and Society* (London and New York: Methuen, 1987), p. 1.
7. Ibid., pp. 6–7.
8. Ibid., p. 7.
9. Walter Fernandes, Geeta Menon, and Philip Viegas, *Forests, Environment and Tribal Economy* (New Delhi: Indian Social Institute, 1988), pp. 2–3.
10. Noted in a presentation by the Inspector General of Forests, Mr. C.D. Pandeya, at the National Workshop on Joint Forest Management, Surajkund, August 1992.
11. Society for the Promotion of Wastelands Development (1984).

12. Madhav Gadgil, Nerendra Prasad, and K.M. Hegde, 'Whither Environmental Activism', *Deccan Herald* (Bangalore), 19 February 1984.

13. Michael R. Miller, 'Mycorrhizae and Succession,' in W. Jordan, M. Gilpin, and J. Aber (eds), *Restoration Ecology: A Synthetic Approach to Ecological Research* (Cambridge: Cambridge University Press, 1987), p. 205.

14. Ibid., p. 209.

15. Nicholas Malajczuk, Norman Jones, and Constance Neely, *The Importance of Mycorrhiza to Forest Trees*, Land Resources Series no. 2 (Washington, D.C.: The World Bank, n.d.), pp. 2–5.

16. Norman Meyers, *Deforestation Rates in Tropical Forests and Their Climatic Implications* (London: Friends of the Earth, December 1989), p. 33.

17. World Bank, 'India's Environment: A Strategy for World Bank Assistance' (New Delhi: India Department, World Bank, March 1989), p. 4.

18. According to India's National Forestry Policy Act of 1952, one-third of the nation's land area was to be brought under forest cover.

19. D.M. Chandrashekhar, B.V. Krishna Murti, and S.R. Ramaswamy, 'Social Forestry in Karnataka: An Impact Analysis', *Economic and Political Weekly*, 22(24): 935, 13 June 1987.

20. N.C. Saxena, 'Farm Forestry in Different Agro-Ecological Regions of India' (paper prepared for the International Workshop on India's Forest Management and Ecological Revival, New Delhi, 10–12 February 1994), p. 5.

21. World Bank, 'Uttar Pradesh Social Forestry Project' (Credit 925–IN), Project Completion Report, 22 May 1989.

22. See N.C. Saxena, 'Farm Forestry in Different Agro-Ecological Regions of India', p. 3.

23. Government of Gujarat, 'Lifting of Seedlings in Farm Forestry Programme', Gandhinagar, Forest Department, Government of Gujarat.

24. IIPO, 'A Report on the Survival Rates of Trees: 1983–84 to 1987–88' (New Delhi: Indian Institute of Public Opinions, 1991).

25. Mandred Seebauer (ed.), 'Study of Social Forestry Programmes in India: Draft Final Report' for the National Wastelands Development Board and the World Bank, p. 1.

26. See N.C. Saxena, 'Farm Forestry in Different Agro-Ecological Regions of India', p. 4.

27. Subhabrata Palit, 'The Future of Indian Forest Management: Into the Twenty-First Century,' Joint Forest Management Working Paper no. 15 (New Delhi: The Ford Foundation, December 1993), p. 3.

28. WWF and SPWD, 'Case Study on Participatory Forest Management in West Bengal' (New Delhi: World Wide Fund for Nature-India and the Society for the Promotion of Wastelands Development, 1994), pp. 17–18.

29. Manjul Bajaj, 'An Examination of the Potential for Investment in Natural Regeneration of Degraded Forests with Community Participation', *Wasteland News*, August–October 1990, pp. 7–22.

30. Ibid., p. 18.

31. N.C. Saxena, T. Shah, and R. Chambers, *To the Hands of the Poor* (New Delhi: Oxford University Press, 1989).

32. P. Guhatakurta, 'The World Bank in the Forestry Sector: Policies and Issues in India', *Wasteland News*, August–October 1993, p. 16.

33. A.N. Chaturvedi, 'Management of Secondary Forests', *Wasteland News*, August–October 1992, pp. 39–44.

34. M. Poffenberger, B. McGean, A. Khare, and J. Campbell, *Field Methods Manual: vol. 2, Community Forest Economy and PRA Methods in South Gujarat, India* (New Delhi: Joint Forest Management Support Programme, 1992), pp. 47–66.

35. V.K. Bahuguna, Vinay Luthra, and B.M.S. Rathor, 'Collective Forest Management in India', *Ambio*, 23(4–5): 269–73, July 1994.

36. Information from conversations with Deep Naryan Pandey, district forest officer, Udaipur South Division, Rajasthan, February 1994.

37. Deb Debal, 'An Estimation of the Possible Effect of Community Leaf Harvesting on the Productivity of Sal' (Calcutta: IBRAD, 1990).

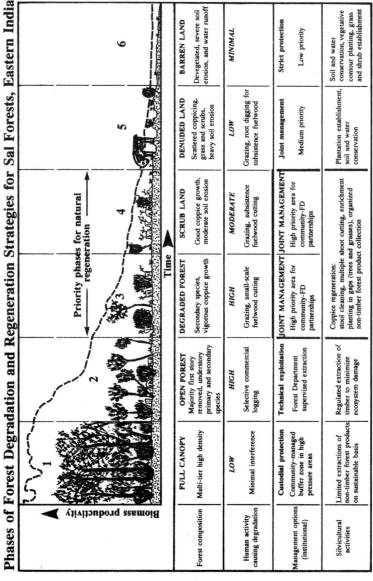

FIGURE 10.1 Common stages of forest degradation

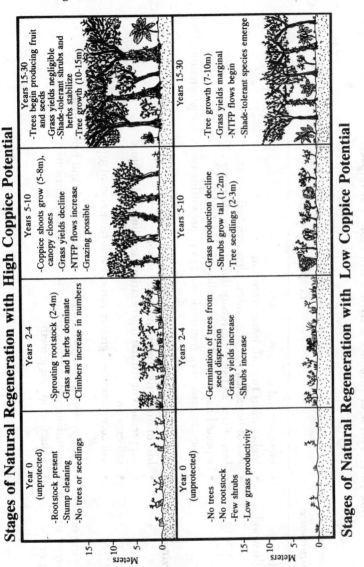

FIGURE 10.2 Typical stages of natural regeneration

Box 10.1: Forest disturbance activities and ecological impacts

Forest Disturbance	Ecological Impacts
Non-sustainable extraction of firewood	• increased soil erosion • higher water run-off, lower retention • loss of branches, leaf area index, photosynthetic parts • loss of tree and shrub diversity due to excessive removal • repeated harvest leads to mortality of tree/shrub • in severe cases even rootstock is dug up, preventing coppice regeneration
Excessive grazing	• loss of diversity and palatable species — emergence of non-palatable species • loss of protective vegetative cover to soil • suppression of natural regeneration: seedling and coppice • soil erosion • soil compaction — greater run-off and reduction of groundwater recharge
Excessive lopping of tree leaves for manure or plate making or for fodder	• canopy opening — invasion of non-native shrub and herb species • disturbance of nutrient cycling — loss of soil nutrients • reduction in tree growth

Non-sustainable legal and illegal timber felling	• loss of tree and tree species and companion species
	• canopy opening
	• more damage to plants due to felling, shaping, and transportation
	• increased soil erosion
	• increased water run-off, decreased waterholding capacity and slow discharge ability
Forest fire or grassland fire	• damage to natural regeneration
	• higher and more tender grass growth for next season
	• fire-resistant tree species survive and others induced to germinate
	• exposure of soil to high intensive premonsoon rains
	• increased soil erosion
	• increased nutrients temporarily
	• decrease in mycorrhiza due to high soil temperature and loss of food supply

Box 10.2: Ecological impacts of conventional block plantations

Parameters	Ecological Impacts
Tree species composition and loss of biodiversity	• loss of local tree diversity in forests, farms, commons • dominance of a few exotic tree species • vulnerability of monoculture to pests, disease, drought, etc. • loss of micro-habitats leading to decline in wildlife diversity • high density suppresses ground vegetation • loss of companion species for climbers, herbs • loss of multi-tier micro-habitats in degraded forests, farms, commons, for birds, insects, reptiles, micro-organisms, etc. • loss of diverse biomass and continuous product flow to humans: food, fodder, raw materials for artisans, manure, oilseeds, etc. • loss of soil quality (micro-organisms, nutrients, etc.)
Sustainability	• repeated short rotational harvests lead to reduction in soil fertility; monocultures are vulnerable to pests, disease attacks
Grass growth	• high density monocultures suppress grass growth

Productivity	• low productivity of plantations in the absence of appropriate soil and water conservation measures and high-quality seeds or coppice
Soil nutrient status	• reduction in soil nutrient status compared to mixed cultures; conservation or primary forests • soil erosion — loss of clay, silt, and humus in topsoil
Groundwater recharge and use	• generally known to be low in plantations compared to mixed cultures or natural forests • surface run-off high • exotic monocultures tend to be more water-demanding

Box 10.3: A Case from South Rajasthan

The Aravalis are India's oldest mountain range. They provide a natural barrier to the expansion of the Great Thar desert into eastern Rajasthan and the fertile Indo-Gangetic plain. Yet, deforestation in this critical environment is perhaps as severe as any place in India. Recent satellite images of the northern Aravalis of Haryana indicate that 80 per cent of the land area is badly denuded, while only 5 per cent of the region maintains some forest cover, generally in poor to moderate condition.[1] The deterioration of vegetative cover in the Aravalis has 'resulted in frequent incidence of drought, famine, cloud burst, lowering the water table, shrinkage of natural forest and pasture lands, and increasing socioeconomic stress on the life of the tribals and hill people.'[2]

During the 1970s, the Rajasthan Forest Department, responsible for sustainable management of the Aravalis, leased extensive areas for commercial logging. After the central government placed restrictions on timber exploitation in the early 1980s, much of the commercial felling was halted. With the assistance of the central government and foreign donor agencies, extensive reforestation schemes were implemented, but these have had little impact in stemming deforestation and ecological degradation. This experience suggests that past and current assumptions and strategies to stabilize the Aravalis need serious review. A case from the forests of southern Rajasthan provides a compelling illustration of the dilemma.

In southern Rajasthan, a divisional forest officer (DFO) notes that the forest in his division covers more than 260,000 hectares, much of which is already severely denuded. Logging carried out until the mid-1980s removed much of the old growth forest in the division. Tribal, Rajput, and Gujar villagers, seeing the logging, also began heavily exploiting the forests through overgrazing and fuelwood cutting, and by setting fires to facilitate mahua flower and honey collection.

Currently, the DFO estimates that at least 140,,000 hectares of the forest area in his division are experiencing serious devegetation and soil erosion through uncontrolled use. Each year these ecosystems lose biomass, biodiversity, and topsoil, which progressively undermine their resilience to re-establish vegetative cover.

The DFO, with a small staff of range and beat officers, is unable to protect the vast forest area under his jurisdiction. Instead, much of his staff's time is devoted to planting a few thousand hectares of highly degraded land, while more than one-half of forest lands under his administration are rapidly degrading. Despite the innovative technical strategies being employed in the small project sites, requirements for budgeting, planning, nursery establishment, protection, and monitoring are staff intensive. As a result, donor support is drawing staff attention away from resolving the resource conflicts that are driving forest degradation on much of the forest lands. If such capital intensive, technically oriented social forestry strategies continue for another ten years in south Udaipur, while a few thousand hectares might be rehabilted, the vast majority of divisional forest lands could experience further degradation and lose much of their regenerative potential.

1   J.P.L. Srivastava and R.N. Kaul, *Greening of the Common Lands in the Aravalis*, Government of Haryana Forest Department: Gurgaon, p. 5, 1994.
2. Ibid., citing Professor Dhabriya, *The Pioneer*, Press Trust of India, 16 December 1993.

Conclusion

Mark Poffenberger ■ Ajit Banerjee

Introduction

Forest management systems in India, as in many of the world's nations, appear to be passing through a transitional phase. Throughout the colonial and post-Independence era, forest policy has emphasized commercial timber use. However, over the past decade, these policies have been questioned and disputed. The 1980 Forest Conservation Act placed stricter controls on timber extraction, while the 1988 National Forest Policy reversed priorities to emphasize ecological balance and conservation. Medium- and large-scale industries can no longer depend on natural forests for unlimited and highly subsidized raw materials to supply their packaging, pulp, paper, and other needs. Instead, the government hopes that these requirements will be met from contracted plantations on private lands or from imports. Such changes reflect important policy shifts for India's natural forests. As in many countries, these new policy directions have been resisted by the commercial sector. In remote areas where valuable timber remains unexploited, business entrepreneurs, with support from the politically powerful, still apply uncompromising pressures on forest department staff to continue extractive activities or look the other way. Yet, with so much of India's natural forest land too degraded to be used profitably, larger businesses are also seeking alternative sources of raw materials.

New Directions in Forest Management

Where rich forests once stood, now there are rock-strewn scrub lands. Timber entrepreneurs have left these lands, providing opportunities for communities to protect and biologically renew them. As democratization grows across the rural countryside, politicians are being forced to be more accountable to their increasingly vocal rural constituencies. Based on their field experiences, professional foresters, too, have sensed the futility of attempting to police vast tracts of forest while in direct conflict with the communities that inhabit, surround, and depend upon them.

These social and ecological changes are undermining the powerful alliances that historically influenced forest use and encouraged unsustainable commercial extraction. Growing rural communities, confronted by increasing forest resource scarcities, are beginning to impose strict controls on access. These remarkable behavioural changes, characterized by consensual community decision-making and group-imposed regulations, may presage a new era of sustainable management. The striking natural resilience of many Indian forest ecosystems to recover rapidly from past disturbances, is also being better understood and supported as a strategy for environmental restoration.

Over the past five years, state orders and a national resolution were formulated to create a 'massive people's movement' to protect and restore forests. Donor and government agencies began shifting their heavy investments in social forestry from private and communal plantations to an alternative approach — the joint forest management (JFM) programmes — involving community protection of public lands. This transformation in strategy has far-reaching implications.

Social forestry programmes largely helped to organize and subsidize the establishment of fast-growing, exotic tree plantations. Millions of dollars in foreign loan funds and billions of rupees in government funds have financed the creation of thousands of nurseries, labour for digging pits, and guards for protection. While such social forestry programmes are acclaimed for establishing plantations on several million hectares, the effort has largely failed to motivate and empower communities to strengthen their own resource management systems. Nor did it improve the condition of the nation's natural forests. In essence, the social forestry programmes of the 1970s and 1980s channelled materials and capital inputs to rural areas to encourage fast-growing tree plantations on private and common lands. While some farmers benefited substantially from these projects, many rural forest communities and tribal villages found them of little help.

JFM strategies, in contrast, require the government to empower communities, allowing villagers to protect natural forests that fall within the public domain. The JFM approach necessitates management decentralization by establishing effective communication channels, agreements regarding rights and responsibilities, dispute arbitration mechanisms, and trust between foresters and villagers. Enhancing the tenure security and authority of communities over forest resources is central to the success of this strategy. Ultimately, JFM represents a historic reversal by allowing decentralization of controls over vast tracts of public forest lands.

The JFM philosophy stands in stark contrast to conventional social forestry projects, which have been based on traditional development paradigms. These have primarily assumed that capital and technical inputs could alleviate poverty, stabilize forest disturbance patterns, and improve productivity of the forest resource base. Since socio-political change inherent in JFM radically deviates from traditional development strategies, it raises important questions regarding the capacity of government and donor agencies to provide support and facilitation that are responsive to the process.

The role of government and donor agencies in supporting new community resource management initiatives is at present poorly defined or tested, and hence, problematic. For several decades, the central government and multilateral and bilateral donors have contributed huge sums to finance India's social forestry programmes. Optimistic government officials claim that 17 million hectares were planted, with a survival rate of 60 per cent. The actual level of field achievement remains debatable. In one case, researchers found that over-reporting of tree planting exceeded 1000 per cent.[1] Foresters from Bolangir District in Orissa reported survival rates as low as 5 to 20 per cent.[2] Despite over-reporting and low survival rates, the massive tree-planting campaign of the past twenty years has stimulated awareness, private farm forestry, and the establishment of millions of trees along canals, roads, and community commons. These initiatives, however, did not attempt to address more fundamental problems of the degradation of public forest lands and the protection and management of natural forests.

Social forestry projects have been criticized for their inability to build community forest management institutional capacity, to reform policies that excluded user groups, or to generally improve the ecological state of natural forests. Increasingly, development banks and other assistance agencies and foundations are interested in funding JFM activities. Recent loans to the government of India for JFM programmes, either finalized or under negotiation, already total several hundred million dollars. However, because JFM represents a very different approach to forest management, it is unclear whether these new initiatives will follow the pattern of conventional social forestry assistance. For donors, a basic constraint is that JFM is not dependent upon, nor designed to absorb, large amounts of capital for employment schemes or technical inputs. Rather, its success is based on reducing conflicts, opening communications between forest communities and the forest department, and counting on voluntary community protection to enhance low or no-cost secondary forest succession. To that end, JFM involves a process of institutional capacity-building, both within the state agency and the community.

Successful JFM programmes need investments in 'software', such as reorientation and gender sensitization in training, the social management of effective community user groups, and the restructuring of policies and procedures that provide flexibility to respond to the socio-cultural diversity so prevalent in rural India. Non-conventional, applied, multi-disciplinary research is the key to the new learning that must continuously inform how government agencies and forest villages can work together more effectively. Much remains to be learned regarding the strengths and weaknesses of community management institutions and the types of support required to enhance their capacity. In technical areas as well, thorough studies of the varied patterns of natural regeneration and flows of important non-timber forest products are essential. Research will need to be more field-oriented, allowing rural people to conduct a wide range of trials, sometimes with assistance from foresters, university researchers, and NGO staff.

Ultimately, the central government and donor agencies must work with state forest departments to move away from achieving physical targets towards developing new skills and capabilities both within forest departments and village user groups. Foresters will need to shift their attention from managing budgets and paperwork, to monitoring field-level forest conditions and talking with villagers. Ultimately, success or failure is not determined by what happens in the budget, but by what happens on the land. More than large infusions of cash, these objectives require improved information, communication, understanding and tolerance, and flexibility. Many state forest departments are still exploring how they can most effectively support grassroots forest protection initiatives. These emerging support programmes have limited fiscal absorptive capacity. For that reason, flooding nascent JFM programmes with an over-abundance of development funds could undermine the efforts of state forest departments to establish co-operative relationships with forest villages. Informal village groups attempting to protect forest resources could be confused by large project budgets. It may, therefore, be more prudent for donors and the government to finance small-scale training, extension, and research projects, thereby allowing both themselves and state agencies to learn more about strategies that support community forest management groups.

Enhanced understanding about how current JFM policies and programmes will affect forest management control in the field is a critical need at this juncture. Current research suggests that community efforts to protect forests in eastern India are largely a spontaneous process of socio-political change, which can effectively expand in the absence of opposition by forest departments. They appear to have emerged from a growing awareness of forest scarcities and dynamic local leadership. Forest departments can help to create these conditions by playing a more proactive, supportive role in enhancing tenurial security and providing technical guidance to community management groups. Given the immense scope for more intensive controls in the forests of the subcontinent, the significance of JFM as a strategy to promote public land reform is staggering. Establishing the political will to transform forest department policies, procedures, and attitudes, which would accelerate this process into the twenty-first century, remains an open-ended challenge.

Predicting the Future of Indian Forest Management

This book has documented past and emerging patterns of forest management, providing a basis to look into the future. Given the current directions of change, it is possible to speculate what may occur in the coming thirty years. While trends of the past decades have given rise to ominous predictions, the emergence of grass-roots environmental movements and supportive policies in recent years provide new hope regarding more sustainable forest management. Ajit Banerjee, who has spent more than thirty years studying Indian forestry, suggests that Indian forestry is entering a transition period extending from 1995 through 2010, followed by a stabilization between 2011 and 2025. Banerjee predicts that India's forest will have a 'rough ride' in the near future, suffering further damage in the coming fifteen years. During 1995–2011, more forests will be cut down and many more will be degraded, especially in areas where the forest is still abundant such as in north-east India, the Himalayas, Madhya Pradesh, in southern and western Orissa, and south Bihar. The primary agents of degradation will be local people practising shifting cultivation, poor villagers collecting fuelwood for local sale, and unscrupulous timber merchants encouraging poorer people to fell forests and convert land to agriculture, which local elites will later claim. Yet, there is hope that further forest degradation will gradually slow as increasing areas come under the control of communities. The stabilization process will come about more quickly in eastern India where local controls are advancing rapidly, and only later in the north-east and the west. During the entire period, large commercial forests will be increasingly degraded. This will happen because of lack of regeneration after selective felling by forest corporations, forest fires, hacking, and excessive grazing.

Banerjee predicts that in the coming decade, forest departments will continue to expend enormous sums on planting a few fast-growing species over large tracts; however, the quality and survival of plantations will continue to be poor and receive little support from the local people. Farm forestry, however, will successfully expand, with farmers not only growing fuel and pulpwood species, but also establishing commercial species such as teak, *champ*, *alstensa*, and *semul* on private lands.

During the first decade of the next millennium, forest departments may continue to reorganize to strengthen their custodial position. While policies to devolve management rights and responsibilities to community groups will gain greater political support, such initiatives will be strongly resisted, especially in areas where forests retain their commercial value. Banerjee believes that only by the year 2025, will forest departments have fully transferred reserved and protected forests to local people for management. By then, however, forest departments will have been transformed into technical extension agencies, assisting communities with support services. Consequently, while the forest department will continue to be an important actor, it will be less as an owner and more as a partner in forest rehabilitation.

With these institutional changes, management systems will be drastically altered. Large blocks of forest — with the possible exception of national parks and biodiversity reserves — will be divided into small, contiguous blocks with each block being managed by separate communities. Communities will play a primary role in forest planning, harvesting, transport, and pricing. Greater emphasis will be placed on producing non-timber forest products. Communities will reduce the level of timber felling and rely less on heavy equipment, transporting timber with human and animal power, and increasingly mill lumber on site. The conversion will not only add value to the timber, but the waste will be retained in the forest, enriching the soil. While communities will receive assistance from forest departments and NGOs in pricing and marketing, local user groups will become the sole arbitrators in sales and retain forest-generated income.

The structure and composition of natural forest ecosystems will change during the coming thirty years. Although the boundaries of the well-stocked forest tracts will remain, the number of trees per hectare will fall, especially of commercially valuable species. This is likely to affect five to ten million hectares of high forests in the Himalayas, parts of Madhya Pradesh and Andhra Pradesh, and the western and eastern Ghats. Biodiversity will also decrease, with increasing numbers of damaged and diseased trees, dense weeds, and climbers remaining. In the mountains, some forests will be totally denuded, exacerbating erosion, landslides, and downstream flooding.

On the positive side, however, much of the degraded tracts will begin to regenerate into dense, low-profile, two-storey coppice forests. By the year 2025, more than 10 to 15 million hectares of degraded forests will be showing healthy signs of regeneration, primarily in central India and the Deccan. Under local protection, these regenerating forests will have better stocking, improved biodiversity, and yield a steady flow of non-timber products sustainably harvested. Grazing controls and better stocking will conserve moisture and reduce erosion, improving soil fertility and agricultural production.

Small tree farms will also become far more extensive throughout Uttar Pradesh, Haryana, Punjab, Jammu and Kashmir, Karnataka, Tamilnadu, and Gujarat. The farms will be predominantly stocked with eucalyptus, acacias, casuarina, poplar, and teak. They will, to a great extent, meet requirements for plywood and pulp, packing cases, and furniture, as well as local needs for fuelwood and construction timber.

Required Changes

Although conventional paradigm contends that economic development in poor nations will eventually renew and protect the forests, India's experience suggests that socio-political strategies involving community empowerment and attitudinal and institutional reforms are a prerequisite to achieving stabilization of forests as renewable resources. Necessary steps in the process involve developing applied research and support programmes that can monitor local strategies to improve forest management at the field level, develop improved communication between forest communities and forest department staff, mediate conflicts, and negotiate consensual management agreements. Monitoring changes in vegetation through field-level information collected by villagers and foresters, as well as through remotely-sensed data, can help determine whether forest ecosystems are regenerating or continuing to degrade.

To achieve a sustainable system of co-management, a new and unprecedented partnership between the government and rural forest communities must be initiated. This process is often catalyzed and facilitated through local leaders, forestry field staff, and small NGOs. Through trial-and-error experimentation and close monitoring, the nature and variations of these partnerships will continue to unfold in the years to come. Despite the need to maintain flexibility and diversity in such newly forged alliances, certain major changes in the structure and orientation of the two primary institutional actors in JFM, namely the forest department and community user groups, are an essential prerequisite. The reorientation will involve a major redefinition of roles in which power and authority are to be shared. Foresters face new challenges in working with communities to allocate territorial management responsibilities, demarcate boundaries, register user groups, develop and extend new management techniques, while providing accurate marketing information. All actors will need to gain an appreciation for each other's attitudes and limitations, while learning to listen to one another to achieve compromise and consensus which is an integral part of the management transition.

The transformation in management is part of a historic process of social and political change. It reflects a reversal of power from the state to the community. Forest departments, proud of their century-old tradition as resource custodians, may find the process painful; delegating responsibilities to rural communities may be seen as a sign of lost authority. As planners and forest administrators confront these decisions and policy changes in the coming decades, they may find it helpful to reflect on the words of Gandhiji:

I will give you a talisman. Whenever you are in doubt, or when the self becomes too much with you apply the following test. Recall the face of the poorest and weakest man whom you may have seen and ask yourself, if the step you contemplate is going to be of any use to him. Will he gain anything by it? Will it restore to him a control over his own life and destiny?[3]

Notes

1. R.V. Singh, 'Timber Demand in India: Prospects for Future Supply and Substitution', in Anil Agarwal (ed.), *The Price of Forests* (New Delhi: Centre for Science and Environment, 1992), p. 70.
2. Amit Mitra and Kanti Kumar, 'Death by Starvation', *Down to Earth*, 15 June 1993, p. 35.
3. Cited by Kamla Chowdhry, 'Social Forestry: Roots and Failure', *Mainstream*, 1 July 1989, p. 34.

Glossary

ashram	place of retreat, meditation, and prayer
baan	rope-making
bhabbar	forest grass
bidis	Indian cheroots
bigha	an Indian land measurement unit, approximately one-third of an acre
chaur	robber, an outlandish or wild person
dati	seasonal harvesting fee
diku	people of the plains
Gaikwad	Maharaja of Baroda
ghatwal	chief
gram sabha	general body meeting
gram	village
jhum/kumri	long-term rotational shifting agriculture
juni gaothan	old villages
Lo Bir	Santhal hunt council
mahajans	moneylenders
mahasangh	co-ordinating body
mandal	management committee
mukhiya	village chief
nistar rights	forest usufructs established during Mughal period or under customary law
paikan	tax-free land for militiamen
paiks	militiamen
panchayat	official system of local/village governance
Panchayati Raj	local governance system

pargana	council
phalia	small hamlet (in western India)
purdah	gender segregation
rajas	rulers
ryots	peasants
sahi	small hamlet (in eastern India)
samiti	council
Sarpanch	panchayat headman
taluka	sub-district
tengga pali	stick rotation system
tola	hamlet
vagh dev	forest spirit/god
vanpal	village forest extension worker
van sathe vikas	development with the forest
zamindar	landlord
zilla parishad	sub-district level representatives

Index

Note: Page numbers of illustrations and maps are in italics.